Generis
PUBLISHING

PLANT BIOTECHNOLOGY

Zareen Narayanan
Bernard R. Glick

Title: **PLANT BIOTECHNOLOGY**

ISBN: 979-8-89966-207-2

Author: Zareen Narayanan, Bernard R. Glick

Cover image: www.pixabay.com

Publisher: Generis Publishing
Online orders: www.generis-publishing.com
Contact email: info@generis-publishing.com

To my parents, husband, and children for their loving support

ZN

To Marcia, Jacob, Sheri, Benji, Aviva and Gabi

BRG

Preface

Since the early days of molecular biotechnology in the 1970s, scientists have sought to use the techniques and approaches of this new discipline to better understand and more effectively utilize plants to deliver a wide range of commercially important products. In this regard, plants have unique properties that can be exploited for practical use. The molecular approaches now common in plant biotechnology have provided a wealth of information about how plants function at the gene level, which has revolutionized the applications of plants in the agriculture, horticulture, forestry, food, vaccine and pharmaceutical industries. In writing this book we have emphasized the major concepts needed to understand breakthroughs in this important field. Throughout we have included numerous diagrams which schematically illustrate all of the key processes discussed. We have chosen not to cover each area in excessive detail. Instead, the focus is on topics that provide examples of important fundamental areas of study useful for an introductory undergraduate course in plant biotechnology. By definition, as with any rapidly moving scientific field, some of the examples that we have chosen are likely to be supplanted as new knowledge is gained. Nevertheless, we believe that this book provides a strong basis for understanding both current and future uses of this technology. We have considered the various plant biotechnological applications and strategies to develop useful transgenic plants, from crop improvement in agriculture to using plants as production systems for pharmaceutical proteins, which should be of interest to postgraduate researchers in the field or scientific workers who need a sound fundamental perspective to this rapidly expanding field. Finally, we have endeavored to write this book using straightforward and jargon-free language so that the concepts and their examples are readily accessible to all readers.

Dr. Zareen Narayanan, University of Washington, Bothell, WA, USA

Dr. Bernard R. Glick, University of Waterloo, Waterloo, ON, Canada

Contents

Chapter 1

Plant Biology

Plants are eukaryotes that manufacture carbohydrates through the process of photosynthesis. The energy driving this synthesis is typically captured from sunlight by chlorophyll molecules (which are responsible for plants' green color). Plant cells are unique in that they include several organelles called plastids that each contain their own DNA and perform specific functions. Thus, amyloplasts are starch-storing plastids, leucoplasts are plastids involved in the synthesis of monoterpenes, chloroplasts are plastids involved in photosynthesis, and etioplasts are plastids whose development from proplastids to chloroplasts has been arrested as a consequence of low light conditions. Plant cells also typically possess a mitochondrion.

1.1 Structural organization in plants

Most higher plants have roots, stems and leaves. The shoot system of a plant includes all the parts of the plant that are above ground, such as the leaves, stems, flowers, and fruits. The root system, on the other hand, consists of all the parts of the plant that are below ground, primarily the roots. (Fig. 1.1). Plants may have parts that differ from each other, but they all have elements that help them perform similar functions.

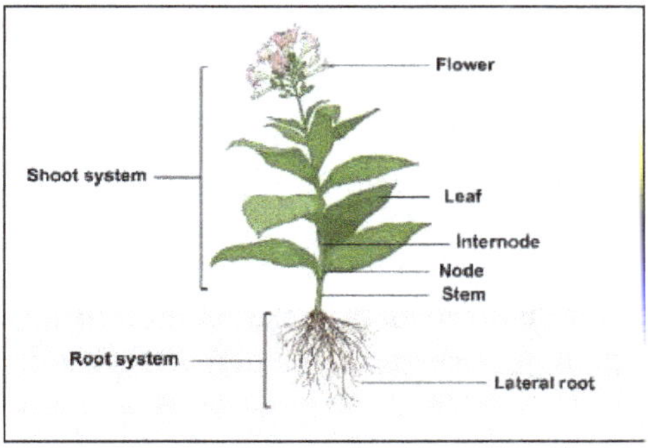

Fig. 1.1 The plant root and shoot system

Roots

Roots are usually the part of a plant that grows under the ground. The primary functions of the roots are to hold a plant in place and to transport nutrients and water from the soil around them and pass it through the stem to the leaves and the flowers. A primary root grows directly downward from the seedling. A primary root develops into a thick taproot that grows down vertically, and from which many smaller lateral roots arise and spread horizontally from the primary root. Tap roots, such as carrots, turnips, and beets, are examples of roots that are modified to store food for the plant (Fig. 1.2).

Fig. 1.2 Storage roots of (a) carrot (b) turnip (c) beet. (photo credit: https://www.inaturalist.org/)

In some plants, the primary root turns into a fibrous root system consisting of many small roots. Grasses such as wheat are examples of plants that contain fibrous root systems. Banyan trees grow aerial roots called prop roots that provide additional support to the plant. Similarly, many mangroves develop aerial roots that extend above the water, allowing the trees to obtain oxygen directly from the air. Plant species living on tree canopies develop special aerial roots called epiphytic roots that help absorb moisture from the air. Some leguminous plants, such as peas and beans, have specialized roots that contain ball shaped structures called nodules. Nitrogen-fixing bacteria live in these nodules. The leguminous plants give the bacteria a home and food in terms of carbon compounds. In return, the bacteria convert atmospheric nitrogen into a form that the plant can use to grow. In addition to this symbiotic relationship, many plants also form mycorrhizal associations between its roots and mycorrhizal fungi.

Root growth occurs at the tips of the roots, specifically the root apical meristem. The tip of the root is covered with a protective cap called the root cap allowing it to penetrate the soil. Cap cells produce a slimy substance called mucigel that aids in lubricating the root allowing it to move easily through the soil thereby facilitating movement of water and minerals from the soil into the plant. Additionally, it attracts beneficial soil bacteria. New cells are produced in the region above the root cap that give rise to the primary body of the plant. This is the region of active cell division. The cells proximal to this region undergo rapid expansion and are responsible for pushing the root cap and the apical tip forward through the soil. This region is called the region of elongation. The cells of the elongation region slowly differentiate into the tissues forming numerous thread-like structures called root hairs which greatly enhance the absorption of water and minerals required during the growth period (Fig. 1.3).

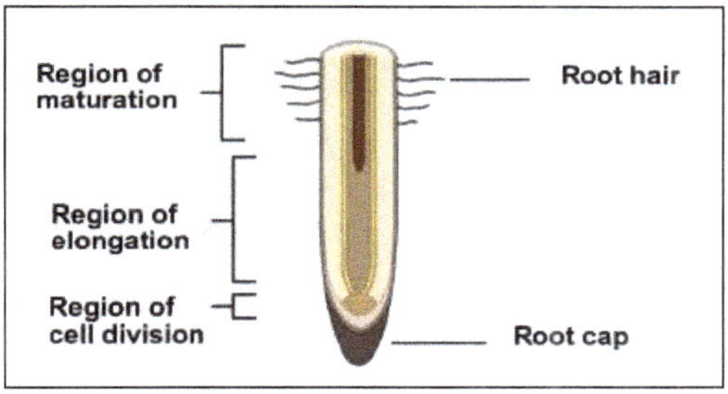

Fig. 1.3 The root zone depicting regions of cell division, enlargement and elongation, and differentiation and maturation. Development of fine, delicate and thread-like structures called root hairs happens in the region of maturation

Stem

The stem provides structure for the plant. Stems do not always look alike. The green shoot of grass is a stem and the trunk of a tree is also a stem. Some stems such as those on a rose bush have thorns on them. The stem of a cactus is usually thick and can store water. The stem contains a vascular system, which transports water and nutrients from the soil throughout the plant. The vascular tissue is a specialized tissue comprising of xylem (for water transport) and phloem (for sugar transport) that form along the length of the stem.

Plant stems, whether above or below ground, have regions called nodes and internodes. Nodes are points on the stem from which a leaf grows. The stem tissue between nodes is called an internode. The stem bears buds, that usually appears at the top of a plant's stem, which may be terminal (appearing at the top) or lateral (appearing in the angle between a leaf and the stem). Buds are sites of new growth in plants that may grow into new shoots or they may develop into flowers. Some stems are modified to store food for the plant and enable it to survive the winter. For instance, tubers, corms, rhizomes and bulbs belong to this category. They perform important functions, for example a taro corm is the swollen underground stem of a taro plant, is a vital food source in many cuisines and can sprout into new plants if cut into sections. Similarly, a potato tuber also stores food for the plant. The modified stems of turmeric and ginger are called rhizomes. In *Arundo donax* (giant reed), a bioenergy plant, the rhizomes store nutrients and can grow into new roots and stalks from their nodes. Fig.1.4 shows examples of some of these below ground modifications. Underground stems in strawberries and some perennial grasses have a compact crown near the soil surface forming new shoots and roots. Stems also modify into runners or stolons, which are horizontal sub-aerial stems that produce roots and shoots along their length leading to the formation of new plants or ramets. This can be seen in plants including strawberries and some types of ivy. A sucker is a modified stem that arises from the base of the plant usually from roots and underground portions of the main stem and emerges into new plants. Bananas and pineapple plants both have suckers.

Fig. 1.4 (a) potato plants form fleshy stem tubers (b) onion plants form bulbs that surround an underground stem that gives rise to new plants (c) giant reeds form a mass of stem called rhizomes that generates multiple plants (d) strawberry plants form stolons on surface of the ground and produce new plants. (Photo credit: https://strawberryplants.org/)

Leaf

A leaf is a flattened, typically green, lateral structure attached to a stem. It is primarily the site of photosynthesis. The leaf blade or the lamina, is the broad, flat part of the leaf. The size and shape of leaf blades among various plants vary significantly. They can be large like the leaves of the banana plant or small like the spines of cacti. The leaf is attached to the stem by a short growth (appendage) called the petiole. If you look closely at a wide leaf, you will notice the veins in it. There is usually a middle vein called the midrib and it branches to produce veins, responsible for transporting water and nutrients as well as for structural support. Sometimes leaves are directly attached to the stem and such leaves are called sessile. Some plants may have small, leaf-like structures called stipules found at the base of the petiole (Fig. 1.5).

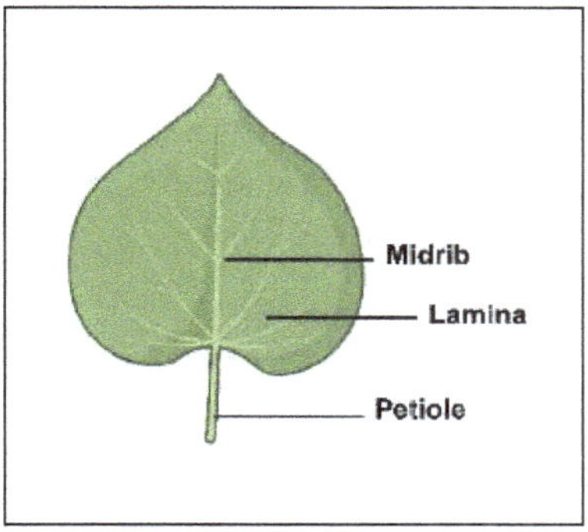

Fig. 1.5 Parts of a leaf

Types of leaves

In a simple leaf, the lamina is undivided. A maple tree leaf, for example, is a single leaf connected to a stem. In a compound leaf, the leaf blade is divided into a few smaller leaves (leaflets), as in the clover tree. Furthermore, if the leaflets are arranged on both sides of a common axis resembling a feather, it is called as pinnately compound or if they emerge at the tip of the petiole, resembling fingers spreading out from a hand, it is called as palmately compound (Fig. 1.6).

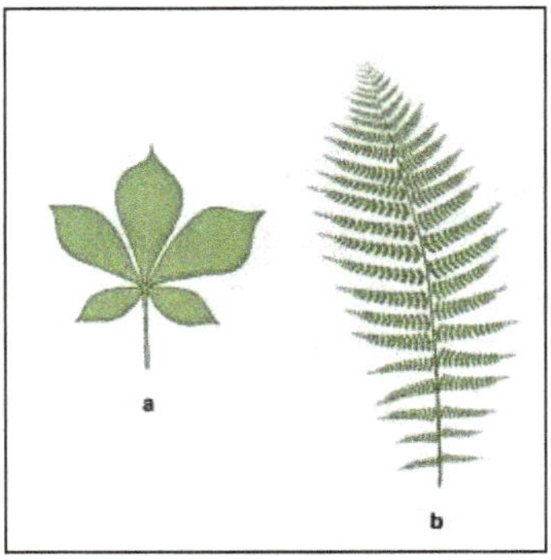

Fig. 1.6 Types of leaves-(a) palmately compound leaf (b) pinnately compound leaf

Phyllotaxy, the positioning of leaves on the stem, is usually of three types. Leaves may be arranged in a spiral pattern, alternation of single leaves, pairs and whorls of more than two leaves at a node. Venation, the pattern of veins on the leaves can be either parallel (when the vein runs parallel) or reticular (when veins form a network). Tulip has leaves with parallel venation whereas maple leaves have reticular venation.

Leaf structure and function

A leaf is made up of several layers (Fig. 1.7). The outermost layer is called the epidermis which protects the leaf. The mesophyll is where photosynthesis takes place. Microscopic structures called stomata in the epidermis allow the exchange of gases through which carbon dioxide enters and oxygen and water exit. Special cells called guard cells regulate the opening and closing of stomata The expansion of the guard cells closes the stomata, while contraction opens it. A waxy layer called cuticle helps trap moisture in the leaf and protect it from diseases.

A plant's vascular system, which includes the xylem and phloem is also present in its leaves. The xylem transports water and minerals to the leaves and the phloem transports food, especially in the form of carbohydrates or sugars, from the leaves to the rest of the plant as a result of photosynthesis.

Leaves look simple but they do amazing work. Leaves are good for plants and people. We eat leaves from a variety of plant species and depend on them for oxygen.

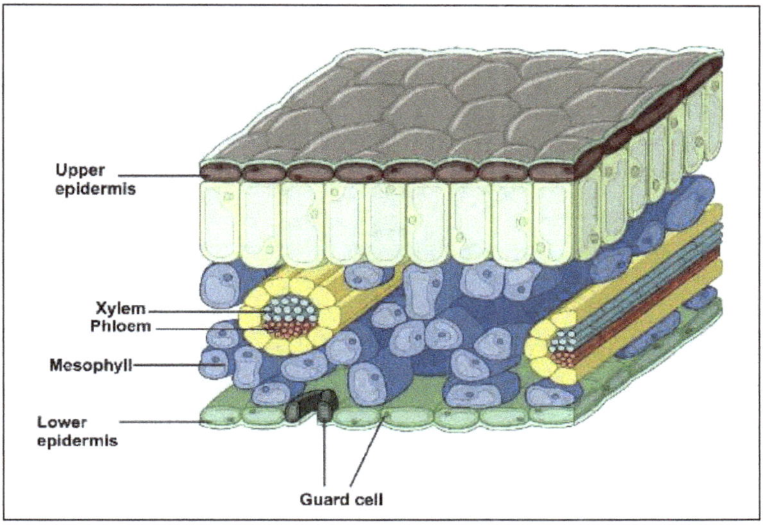

Fig. 1.7 Cross-section of a leaf

Flowers and their parts

Throughout history, humans have cherished flowers for their beauty, their fragrance, their ornamental value, and their importance in social, religious, and cultural traditions. They have been used to convey messages of love, happiness, grief, and more. To a plant biologist, flowers are the part of the plant concerned with sexual reproduction. A flower has several specialized organs that play a crucial role in plant reproduction (Fig. 1.8). Petals, often brightly colored, attract pollinators such as bees, butterflies and birds, facilitating the pollination process. Petals collectively form the corolla and they vary remarkably in their shape and color. Beneath the petals are leaf-like structures called sepals. The male reproductive organs are called stamens. The male gametes or pollen are produced in the anther, a bilobed structure (i.e., two lobed) at the tip of the stamen and connected to the base of the flower by a long, thin filament. The female reproductive organs called pistils typically consists of three parts-the sticky topmost part called the stigma where the pollen grains are trapped; the style is the slender, long column down which the pollen must travel to the ovary and an enlarged basal structure, that is often not visible is the ovary. The ovary contains the female gametes, the ovules, which after fertilization become seeds and the ovary matures into a fruit.

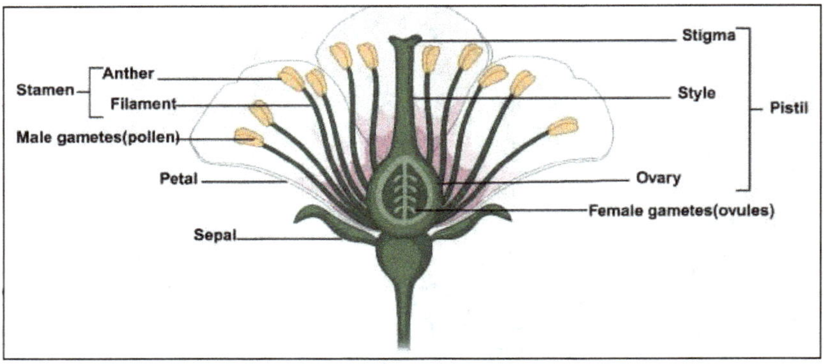

Fig. 1.8 Parts of a flower

Some flowers have all parts of a flower and are called complete flowers. Some don't have sepals, or only contain male or female parts. They are called incomplete flowers. A flower with both male and female parts is called perfect even if it's incomplete. An imperfect flower has either male or female parts. Monoecious plants contain both male and female imperfect flowers. Corn is a classic example of a monoecious plant. It has male flowers (tassels) and female flowers (ears) on the same plant. Dioecious plants will have either male flowers (staminate) or female flowers (pistillate) and can only reproduce when they are in close proximity. Most plants are monoecious.

1.2 Seeds, germination and dispersal

A seed is the mature ovule of a plant and has the essential components necessary for a new plant. Seeds are not just important for the plant but also for humans. A seed is typically composed of a seed coat, an endosperm and an embryo. The function of the seed coat is to protect the seed from adverse conditions, and it can be hard and rigid in some plants (lupine and pea). The endosperm is a storage structure containing important growth substances, including proteins, starch, sugars, which nourish the embryo as it germinates into a new plant. Embryos typically have a small radicle (root part), and a large hypocotyl that becomes the plant's first leaf or leaves (Fig. 1.9).

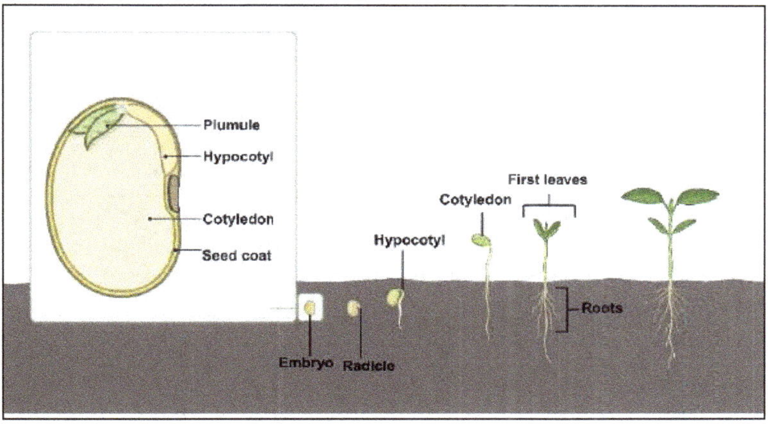

Fig. 1.9 Germination and seedling development in soybean

An embryo with one cotyledon is classified as a monocotyledon, or monocot, which is characteristic of plants such as grasses, lilies, and orchids. Monocots typically exhibit parallel leaf venation, fibrous root systems, and their stem vascular tissue are bundled together. Plants with two cotyledons are called dicotyledons or dicots and they typically have taproots and their vascular system is arranged cylindrically across the diameter of the stem, appearing like layers of growth that turn into wood and bark. See Fig. 1.10 for detailed differences between monocots and dicots. All monocots and dicots are flowering plants or angiosperms. Flowering plants have their seeds hidden within their fruit, although some plants like corn and other grains, have their fruit and seeds fused together. In gymnosperms, the seeds are exposed and this group of plants includes all conifers.

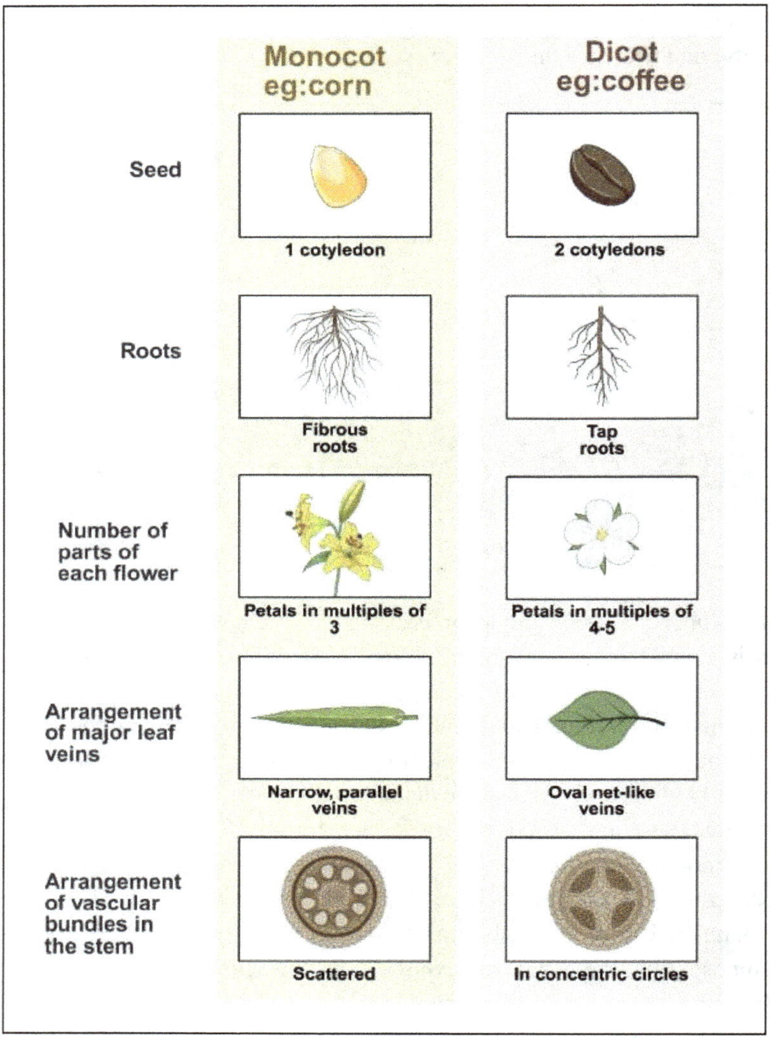

Fig. 1.10 Comparison of monocots and dicots

Seed dispersal methods vary widely among plants, adapting to different environments and ensuring the successful propagation of species. These methods can be categorized into several types, including wind dispersal, where lightweight seeds are blown away by the wind. Plants on a riverbank drop their seeds into the water. Seeds are dispersed by the water. Seeds often do much better when they are carried farther away from their parent plant so that when they grow, they don't crowd each other or have to compete for water or light. Some plants have seeds that are sticky or have burrs that adhere to animal fur and these seeds are carried to new places and then take root and germinate. Often, animals and birds also eliminate the seeds that are in the fruit they eat. Some plants have seedpods that burst open causing the scattering of seeds into the air.

Seeds can survive for a long time before they begin to grow into a plant. Farmers usually purchase the seeds they use to grow crops. Much of today's commodity seeds are proprietary, developed and owned by agricultural companies. These seeds often come with specific traits, such as reduced pesticide use, higher yields, or increased resistance to drought or disease, and therefore are protected by intellectual property laws (see Chapter 6).

1.3 Plant reproduction

Before flowers can produce fruit and seeds, they must be pollinated. During pollination, the pollen, which is produced in the anther, fuses with the ovules inside the flower. Plants are called self-pollinating when the pollen from the anther is deposited on the stigma of the same flower, or another flower on the same plant. This type of plant does not need assistance from pollinators as the pollen can reach the flower's stigma. Cross pollination occurs when the pollen from one flower is transferred to a different flower on the same plant or a different plant via the help of pollinators (Fig. 1.11). Pollinators include wind, waves, human activity and animals such as bees, butterflies, beetles, ants, flies, bats, and hummingbirds.

Fig. 1.11 (a)The flower is being pollinated by the butterfly (b) the grass is pollinated by the wind (photo credit: https://www.inaturalist.org/)

Pollination results in plant fertilization leading to the development of a seed within the plant. The plant then finds the means to disperse seeds and when conditions are favorable, each seed germinates into a new plant. This form of reproduction is sexual, and it promotes genetic variability in the offspring. Sexual reproduction is advantageous for plant survival and adaptation to an ever-changing environment. Sexual reproduction also has several agricultural uses, for example in plant breeding, cross-pollination has been widely utilized through artificial crosses, where breeders intentionally combine desirable traits from different plants. These controlled breeding efforts have been foundational in producing the modern cultivars we rely on today, with enhanced characteristics such as higher yields, improved disease resistance, and better adaptability to various environmental conditions. Blight-resistant corn, rust-resistant wheat, and aphid-resistant alfalfa (*Medicago sativa*) are some notable examples of conventional plant breeding.

Many plants reproduce asexually through vegetative propagation either naturally or by artificial methods. The different types of stem modifications discussed earlier such as rhizomes, stolons, tubers, bulbs and corms facilitate asexual reproduction, resulting in the production of clones. Artificial methods of asexual reproduction can be beneficial for introducing new traits into plants. Plant tissue culture is a technique of culturing plant seeds, organs, explants or tissues on a chemically defined synthetic nutrient media under sterile conditions in a lab (Fig. 1.12). This method rapidly produces clones (identical copies) of desirable plants.

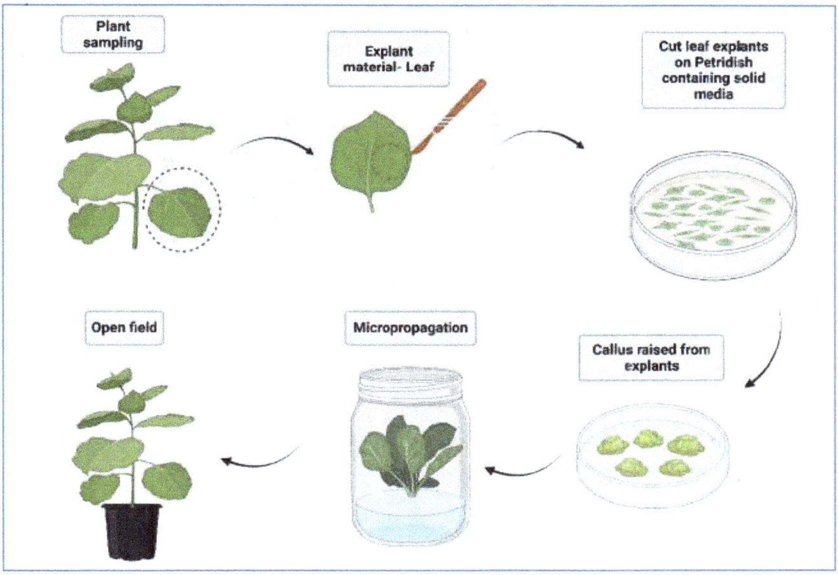

Fig. 1.12 A schematic of the steps involved in plant tissue culture

Grafting is widely used by horticulturists whereby tissues of one plant are joined to the tissue of another plant. For example, small pieces of other plants may be grafted to a root stock of apple tree as a means to generate a multiple fruit tree. Layering involves bending the plant's branch so it touches the ground. After some time, adventitious roots develop and the branch can be disconnected from the parent plant and propagated. Taking cuttings, is quite common for propagating plants such as willows and poplars, whereby a piece of stem is cut and placed in water or soil containing nutrients and the plant will develop new roots (Fig. 1.13).

Fig. 1.13 Cuttings of willow plants placed in conical flasks with water and nutrients (photo credit: from Khan *et al.* 2014)

1.4 Plant cell components

Plastids

Plastids are found in all plant cells and show remarkable diversity both in form, function and the presence of pigments. Chloroplasts are pigmented plastids which carry out photosynthesis (see section 1.6). A chloroplast is characterized by an outer membrane, an inner membrane and membrane structures called thylakoids and the fluid stroma (Fig. 1.14). The thylakoids are arranged in stacks called grana. Plastid outer membrane lipids have been implicated to have a role in precursor targeting and recognition. The membrane system is responsible for trapping the light energy and also for the synthesis of chemical energy in the form of energy-storage molecules-ATP (adenosine triphosphate) and NADPH (nicotinamide adenine dinucleotide phosphate). In the stroma, enzymatic reactions incorporate carbon dioxide into the plants leading to the synthesis of sugar. Chlorophyll, a pigment is found in the thylakoids that absorbs light.

Each plant cell may contain up to 100 chloroplasts. Chloroplasts have their own ribosomes for protein synthesis and a small circular genome (~120-160 kb) that contains the code for chloroplast proteins (e.g., ATP synthase). Whereas some proteins are synthesized within the chloroplast, and are involved mainly in carbon dioxide fixation and other aspects of chloroplast metabolism, many of the chloroplast functions are dependent on the nuclear DNA. Proteins targeted to plastids can be delivered post translationally via an import mechanism (Chapter 3). Several proteins are encoded in the nucleus, synthesized in the cytoplasm and transported into the chloroplast. Moreover, foreign DNA may also be expressed as part of the of the chloroplast genome, enabling production of valuable proteins and agronomic traits (see Chapter 5).

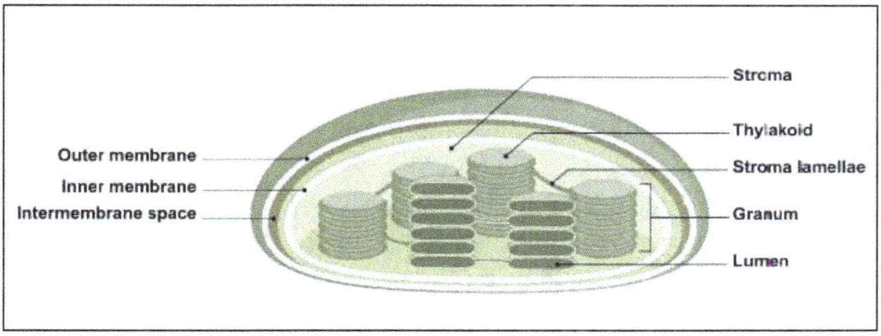

Fig. 1.14 Diagrammatic representation of the 3D structure of a chloroplast

Plastids include other pigmented organelles called chromoplasts which contain carotenes, xanthophylls and anthocyanins that impart color (red, yellow and purple) to certain portions of plants helping in plant reproduction and contributing to visual diversity (Fig. 1.15). Amyloplasts are colorless plastids that store starch. These organelles are present in storage organs like tubers and roots. In some plant tissues, such as potato tubers the amyloplasts can be quite large accumulating substantial amounts of starch. These can be easily viewed under a standard light microscope. Other non-pigmented plastids include elaioplasts that store oils and fats and aleuroplasts (also called proteinoplasts) are protein storing plastids, both occurring in seeds.

Fig. 1.15 β-carotene is responsible for red/orange color in (a) tomato and (b) papaya (c) the purple color in grapes is due to anthocyanins; and (d) xanthophylls impart yellow color to flowers

Nucleus

The most important role of DNA is to carry genes, the instructions to make specific proteins–including details about when, in what types of cells, and in what quantity each protein is to be made. The nucleus is the site where nearly all of the cell's DNA (i.e., outside of the small amount of DNA found in some plastids) is located, replicated and transcribed into RNA. In plants, the nucleus includes the nuclear membrane, nucleoplasm, chromatin and the nucleolus. It is surrounded by a double membrane, the nuclear membrane, which is largely composed of phospholipids. Because it is perforated, the membrane can allow passage of RNA, protein molecules, nucleotides (building blocks for DNA and RNA) and ATP molecules (that provide energy for assembling DNA) in and out of the nucleus. Inside the nuclear membrane is a fluid-filled space called nucleoplasm which contains the nucleolus and chromatin. In an actively dividing cell, the DNA is bundled up tightly in the form of chromosomes, so the cell can move around its DNA more easily. However, when the cell is just functioning, the DNA is very loose within the nucleus, in the form of chromatin. Chromatin is a dynamic functional unit containing histones (specialized protein) and various proteins required for DNA replication. It is best represented by the "beads-on-string" model where the string is DNA, and each bead is a nucleosome unit in which histones bind to and fold the DNA; connected by linker DNA (Fig.1.16). Chromatin modifications (for example, DNA methylation) enable plants to be more competent and is considered an important area in plant molecular biology (see Chapter 3).

Cellular structures called ribosomes are often attached to the outer surface of the nuclear envelope. A functional unit of a ribosome called the 80S subunit consists of a large (60S) and a small (40S) subunit that each contain ribosomal RNA (rRNA) and ribosomal protein. The nucleolus is the site where the two ribosomal subunits are made and are moved out of the nucleus to the cytoplasm, where they join together for protein synthesis.

Proteins play crucial roles in maintaining the structure and function of cells. They serve as enzymes, catalyzing biochemical reactions, and are essential for transporting materials in and out of cells. Proteins also act as receptors for signals, enabling cells to respond to external stimuli. Surface receptors on proteins are vital for cell-cell communication. Additionally, proteins organize molecules within the cell, ensuring proper cellular architecture and function. They regulate how DNA is utilized by the cell, influencing gene expression and cellular activities.

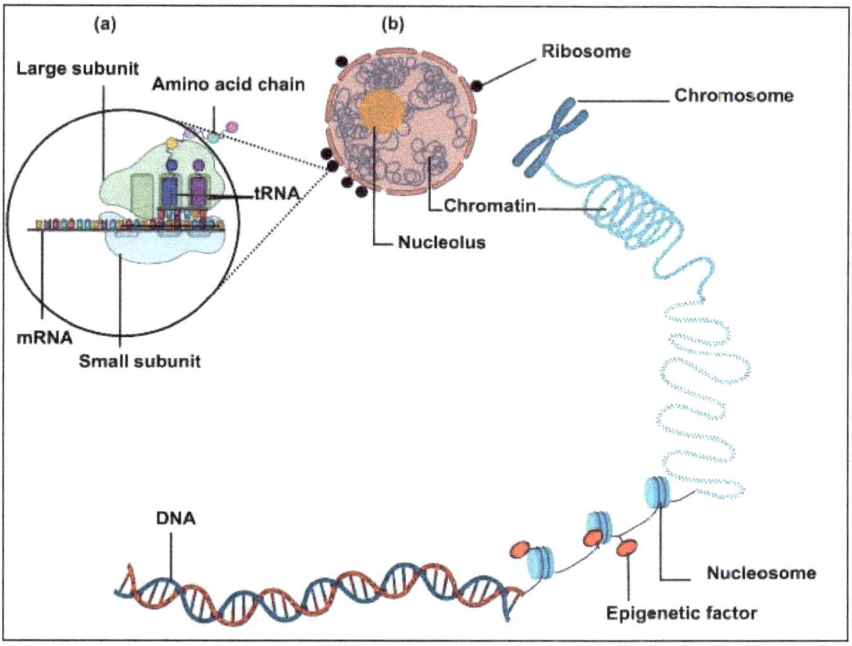

Fig. 1.16 Schematic diagram showing (a) ribosome structure (b) chromatin organization

Mitochondria (mt)

Mitochondria are oval shaped structures with an outer membrane, and an inner membrane, with the two separated by an inner membrane space. The inner membrane of a mitochondria forms folds called "cristae". Inside the cristae is the matrix (Fig. 1.17). Mitochondria are often referred to as the "energy factories" of cells because they generate most of the cells' supply of ATP, which is used as a source of energy for numerous metabolic processes. Key steps of cellular respiration occur in the mitochondria e.g., the Krebs cycle, which takes place in the matrix, and the electron transport chain (ETC) takes place on the cristae.

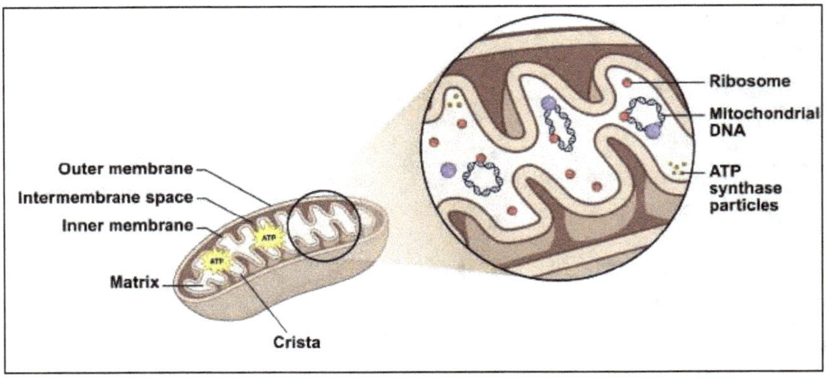

Fig. 1.17 Diagrammatic representation of the 3D structure of mitochondria

Like chloroplasts, mitochondria contain their own DNA and ribosomes for protein synthesis. Typically, mitochondrial DNA is mostly circular however there is a great variation in the size of the mitochondrial DNA among plant species. For example, the mitochondrial genome of *Arabidopsis thaliana* is ~367 kb and that of the flowering plant *Silene conica* is ~11.3 Mb. Unlike chloroplasts, the larger size of the mtDNA provides more space for inserting foreign genes without disrupting essential functions. This may be important in plant genetic engineering where success depends on the integration and expression of target genes without compromising the plant's essential functions. The ribosomes of mitochondria are different from those in the chloroplast and cytoplasm.

Although the mitochondria have the necessary molecular machinery to perform a number of functions including electron transport, ATP synthesis and RNA synthesis, processing and translation, most of the proteins present within the mitochondrion however, are encoded by nuclear DNA, translated in the cytoplasm on 80S ribosomes, and transported into the mitochondria.

Cell walls

The cell wall provides strength and protection to the plant cells. Besides its structural role, the wall is important in several cellular metabolic processes including absorption and secretion of substances, detection of bacterial and fungal pathogens and also in the regulation of plant growth and development. The primary cell wall is mostly composed of cellulose microfibrils that are embedded in a matrix of complex polysaccharides such as hemicellulose and pectin (Fig. 1.18a). The cellulose molecules consist of β-1,4 linked glucose units that form long chains (Fig. 1.18b). Each glucose unit interacts covalently through hydrogen bonds, and is parallel but inverted relative to neighbor, forming a strong network analogous to the structural arrangement of individual threads in a cotton fabric. Cellulose provides most of the tensile strength to the plant cell wall. The hemicellulose is characterized by neutral glycans that bind to the cellulose through hydrogen bonds forming a dense network, making the wall strong and resilient. Pectin is a gel-like polysaccharide whose main component is galacturonic acid, and it helps to hold plant cells together. Woody plants modify their cell walls by depositing lignin in their vascular tissues. Lignin, the second most common material of plant cell walls (e.g., trees), is a complex network of phenolic compounds. It is hard and acts like a "glue" that sticks to the other cell wall components making the wall even more rigid and permanent and more resistant to degradation. Although this is an adaptive strategy that some plants employ, lignin poses a significant challenge to scientists as it is a major obstacle for utilizing cellulose as a biofuel. For improving the efficiency of cellulosic biofuels, it is possible to alter the lignin content of plant cell walls using genetic engineering techniques (see Chapter 6).

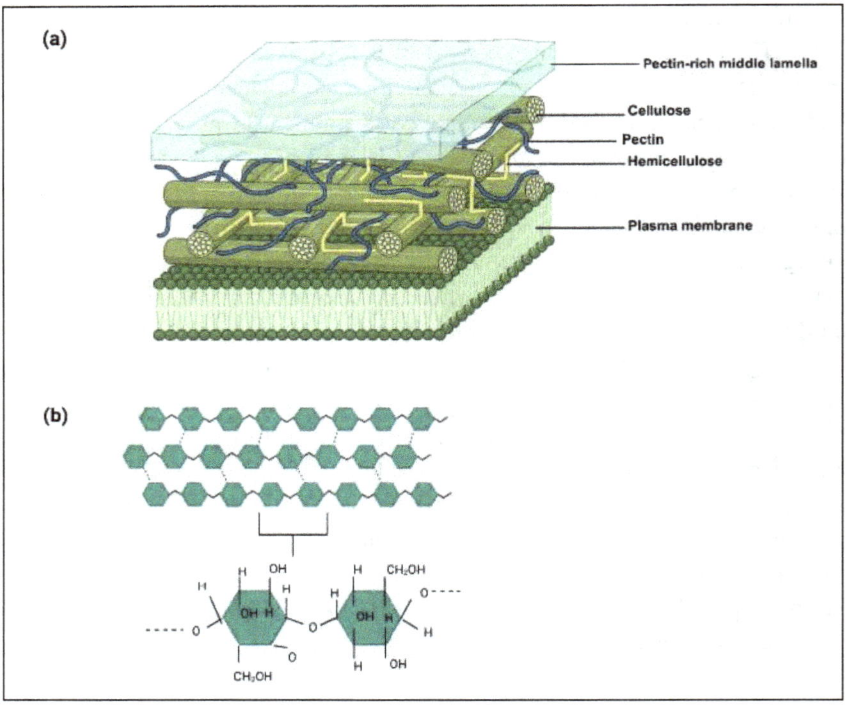

Fig.1.18 A diagram of (a) the structures seen in the primary plant cell wall (b) the structure of a cellulose molecule depicting a long, unbranched chain of β1,4-linked glucose units. Many parallel glucose units are covalently linked to one another to form a microfibril that is mechanically strong

1.5 Plant membranes and receptors

Plant cell membranes are essential for maintaining cellular homeostasis and facilitating communication between the cell and its environment. All of the cellular membranes are similar in structure and are mainly composed of a double layer of *fluid* lipids. The membranes vary in their constituent proteins which float laterally within the lipid bilayer. This ability to move within the membrane forming a changing *mosaic* pattern is how the membrane structure is visualized and is commonly referred to as the fluid mosaic model. The main characteristic feature of the cell membrane is that it is selectively permeable. Membranes, like the plasma membrane selectively permit the passage of some materials while inhibiting the passage of others. Passage of charged and polar molecules through membranes is facilitated through protein channels embedded within the membrane. Some channels exclusively transport water and uncharged molecules (e.g., aquaporins) whereas others transport various metals (e.g., metal transporters). ABC transporters are a type of membrane proteins that use the energy released by hydrolysis of ATP to transport lipid molecules through the protein. Sugars and carbohydrates are moved via anion channels, however high molecular weight compounds such as polysaccharides and larger proteins are transported into balloon-like structures called vesicles. These transmembrane proteins vary from the other type of proteins called the peripheral proteins which are mostly glycoproteins (proteins attached to carbohydrates).

Some of the proteins embedded in the lipid bilayer can act as receptors that detect external signals with instructions for a specific action. Receptor kinases (RKs) and receptor-like proteins (RLPs) are examples of receptors that recognize signaling molecules from potential pathogens. Upon recognition, they undergo a conformational change that allows them to initiate a signaling cascade within the cell. This process called signal transduction activates appropriate cellular responses. The transduced signal is further amplified by secondary messengers to the effector proteins, which then carry out the function. This could mean activation of defense related genes or intra cellular signaling molecules that recruit additional immune responses. The cell thus mounts a defense response to counter the pathogen (Fig. 1.19).

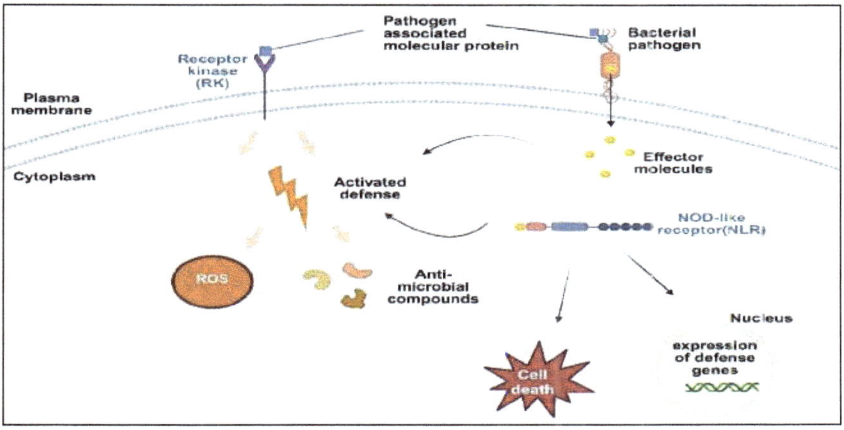

Fig. 1.19 An overview of plant-immune signaling network. Pathogen associated molecular proteins are recognized by receptor kinase triggering immunity. Receptor kinase (RK) activates production of antimicrobial compounds and reactive oxygen species (ROS). Effectors produced

by the bacterial pathogen invading the plant cell are recognized by NOD-like receptors (NLRs) present within the cytoplasm induce defense genes and cause programmed cell death

1.6 Photosynthesis and photorespiration

Photosynthesis is a process by which plants capture energy from sunlight and carbon dioxide from the atmosphere and convert it into sugar, the type of resource that plants need to grow (Fig. 1.20).

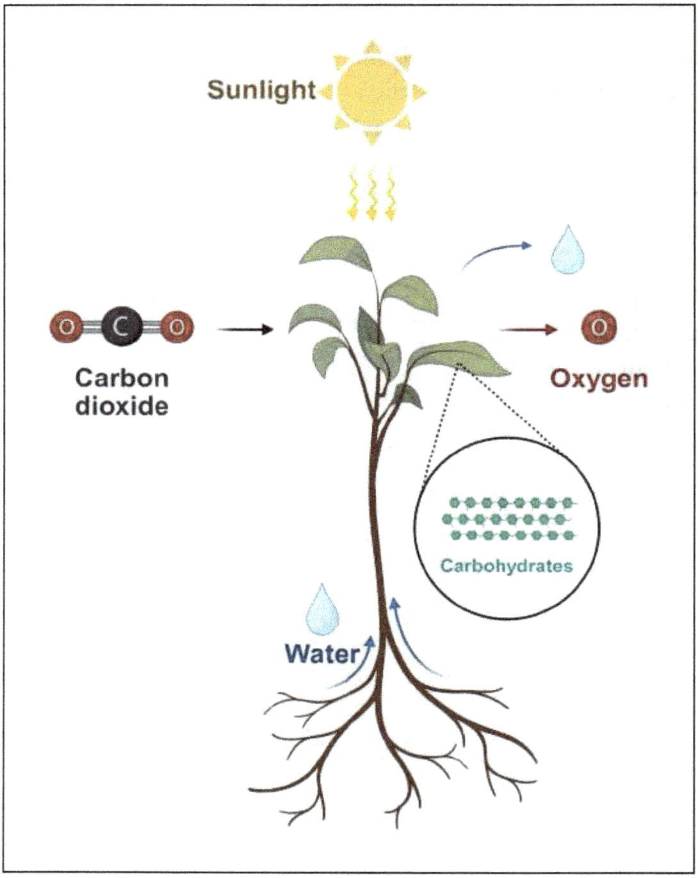

Fig. 1.20 Overview of photosynthesis- plants capture light energy from the sun, and by utilizing carbon dioxide and water, convert it into chemical energy stored in the form of carbohydrates. Oxygen and water are released as byproducts

The following is the general equation of photosynthesis:

$$6CO_2 + 12H_2O \longrightarrow C_6H_{12}O_6 + 6O_2 + 6H_2O$$

Photosynthesis involves two distinct steps- (i) photoreduction or the light dependent reaction. In this step, plants capture light via the light harvesting complex in chlorophyll and temporarily store it as chemical energy in the chemical bonds of ATP and the electron carrier NADPH. In step two, ATP and NADPH are used to drive the processes leading to the synthesis of energy in a form that can be transported and stored, as the carbohydrates- sugar and starch. This process involves a series of events also called the Calvin cycle. The products of Calvin cycle are sucrose, a disaccharide sugar transported from cell to cell and starch, which builds up as an energy reserve that is broken down into sucrose when the cells need it. Oxygen is a byproduct of water molecules splitting in the initial transfer of energy during the first step, and is expelled back into the atmosphere. The Calvin cycle takes place in the stroma of the chloroplasts and consists of three steps: fixation, reduction and regeneration. All of these reactions are catalyzed by enzymes that rearrange the molecules in preparation for the next function. In step 1, the carbon dioxide is added to ribulose 1,5 biphosphate (RuBP), a 5-carbon sugar. RuBP is immediately broken down into two 3-carbon molecules of 3-phosphoglyceric acid (PGA) with the help of the enzyme ribulose-1,5-biphosphate carboxylase-oxygenase or Rubisco (one of the most abundant proteins on earth). In reduction, which is a two-step reaction, PGA is reduced to glyceraldehyde 3-phosphate or 3-phosphoglyceric acid (G3P), using the ATP and NADPH molecules. In the regeneration step, much of the G3P produced is exported from the chloroplast into the cytoplasm and the remaining molecules are rearranged and used to regenerate RuBP, the starting material to which carbon dioxide binds. Thus, the cycle continues (Fig. 1.21).

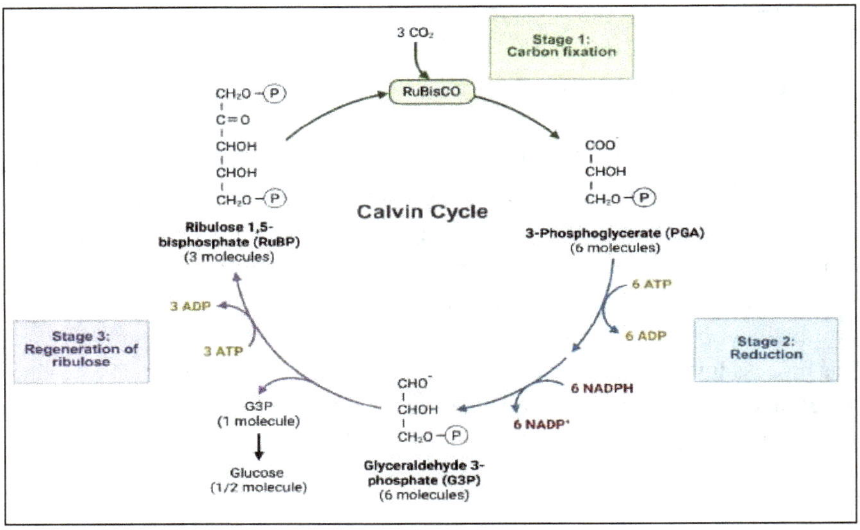

Fig. 1.21 The Calvin cycle

Three turns of Calvin cycle (incorporating a total of 3 atoms of carbon, one at each turn) result in the production of glucose a sugar which is subsequently used by the plant to form a variety of organic compounds required for a plant's structure and function. Most temperate plants such as rice, wheat, soybeans and barley use this pathway. Sometimes, an additional step is added to the Calvin cycle increasing the efficiency of photosynthesis in plants growing in tropical climates (e.g., grasses). This modified pathway of carbon fixation is referred to as the C4 or Hatch-Slack pathway. Similarly, the CAM (crassulacean acid metabolism) pathway is another type of modification made by desert plants (succulents, cacti, agave) in which carbon dioxide is fixed during the night and used to make glucose during the day (Fig. 1.22).

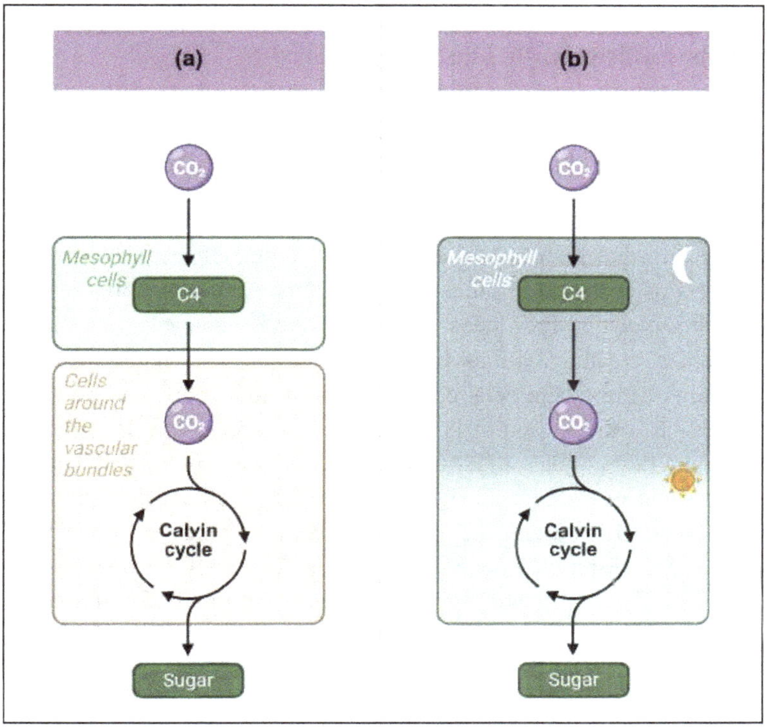

Fig. 1.22 Modified photosynthetic pathways in (a) C4 plants (b) CAM plants

It is important to understand the role of temperature, sunlight, and carbon dioxide concentration in photosynthesis. Usually, plants need ample sunlight but many plants can thrive in areas lacking sunlight or in places with scattered light. Plants mostly require blue and red light for photosynthesis. In the absence of sunlight, special grow lights may be used, for example in greenhouses. Some growing situations lack sufficient sunlight and that's why greenhouses are often equipped with special grow lights for maximizing photosynthesis. Excessively low or high temperatures are detrimental for photosynthesis, and so are extremes of humidity. If there is too much moisture in the air, water droplets will settle on plant leaves and block stomata and there will be no gas exchange. Conversely, not enough water causes the stomata to close up, also preventing the plant from absorbing carbon dioxide. Indeed, the CO_2 concentration also influences the photosynthetic rate. A lack of CO_2 would not only result in a lower photosynthetic yield but the plants grown under these conditions tend to be of inferior quality and strength. Under ideal conditions, a plant can reach its maximum photosynthesis levels, beyond which growth is negatively affected.

Photorespiration also known as cellular respiration is the opposite of photosynthesis. It is a metabolic process by which plants use oxygen to convert glucose (harnessed during photosynthesis) into ATP that can be used by the plant to fuel various biochemical processes. Unlike photosynthesis which occurs in daylight, photorespiration does not stop at night. It occurs partly in the mitochondria of cells and in other parts of plants. Aerobic respiration (requires oxygen) takes place in three steps: glycolysis, the Krebs cycle, and the electron transport chain creating various energy rich molecules, which are eventually converted into a form that cells can use to perform various functions. Aerobic respiration is the primary and most efficient way to produce ATP in plants, while anaerobic respiration (does not require oxygen) is less common and less efficient. Temperature and oxygen levels significantly impact cellular respiration, among other factors. For many plants the optimal temperature range for respiration is 20-30°C, below 5°C respiration becomes very slow, as it does when the temperature is above 40°C. External factors such as drought limit the intake of oxygen from air due to the closing of stomata in the leaves thereby reducing respiration. On the other hand, if plant roots are flooded after heavy rains, the oxygen levels in the soil can drop significantly because water fills the soil pores that normally contain air. This creates hypoxic conditions, and the roots may rely on anaerobic respiration until the soil dries out and the oxygen level is restored. However, prolonged flooding conditions can cause plant death. A plant's developmental stage can also affect respiration, for example, mature fruits or fertilized seeds respire at a low rate. Respiration is minimal or zero in a dormant seed. Additionally, in plant parts where complex metabolic activities occur, respiration rates are generally higher. For instance, respiration is generally greater in leaves than in roots, and younger leaves tend to have higher respiration rates than older leaves.

1.7 Plant hormones

Plant hormones or phytohormones are a group of naturally occurring molecules with diverse structures (Fig. 1.23) that play a vital role in the growth and development of plants. They are synthesized by most plant cells and can either act locally within the cell of origin or be transported to other target cells throughout the plant.

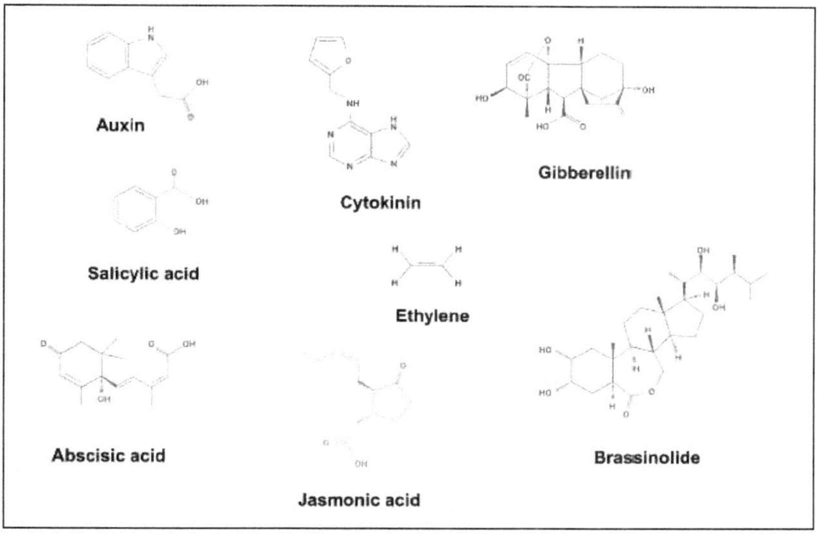

Fig. 1.23 Chemical structures of phytohormones

There are five main classes of plant hormones: auxin, cytokinin, gibberellin, abscisic acid, and ethylene. Other plant hormones include salicylic acid, jasmonic acid and brassinolides. Some of the physiological effects of these hormones are described below.

Auxins

As gardeners and horticulturists often observe, trimming off the tip of a branch can stimulate plant side growth (Fig. 1.24). This phenomenon is called apical dominance, and it occurs when a shoot tip suppresses axillary bud outgrowth and restricts the number of branches. By removal of the stem apex the source of auxin is removed and its inhibitory action on axillary buds is released, leading to side branch outgrowth and bushier plants. The key compound produced in the young and growing leaves of the shoot tip that inhibited axillary buds is auxin. The name auxin comes from the Greek word 'auxein', which means 'to grow'. Auxins play a critical role in a large number of developmental processes in a plant. For example, they initiate production of adventitious and lateral root formation in stems, e.g., in tea plants and apples they induce elongation growth of shoots, induce changes in leaf shape, co-ordinate fruit setting and dropping (e.g., tomato), and induce flowering (e.g., pineapple).

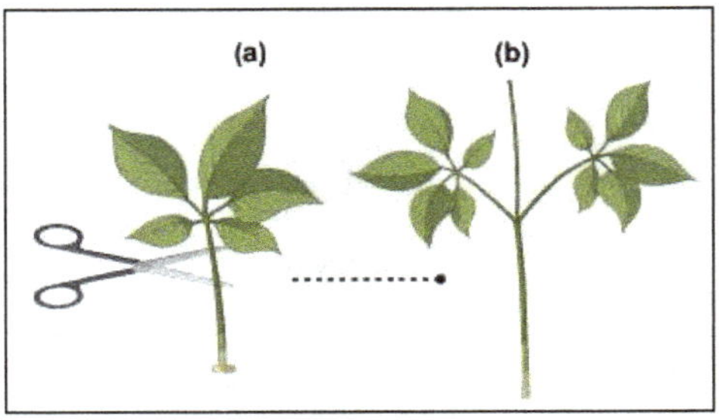

Fig. 1.24 (a) a plant with an intact apical shoot (tip) (b) a plant with tip removed (note the growth of lateral branches on the sides)

The main auxin in plants is indole-3-acetic acid (IAA). Although other compounds with auxin activity, such as indole-3-butyric acid (IBA), phenyl acetic acid, and 4-chloro-IAA, are also present in plants. Naphthalene acetic acid (NAA) and 2,4-dichlorophenoxyacetic acid (2,4-D) are synthetic auxins., Both natural and synthetic auxins have been extensively used in agriculture to increase crop yields. Moreover, exogenously applied phytohormones stimulate the transport of photosynthates, improving the nutrition level of plant tissues ultimately increasing the quality of crops.

Cytokinins

Cytokinins are a group of compounds that regulate cell growth and development. These are synthesized in regions where cell division (cytokinesis) is occurring, for example roots, shoot buds, embryos, fruits, etc. Cytokinins modify apical dominance (caused due to auxins) by initiating bushier lateral bud growth. They are important components regulating the development and function of chloroplasts. Zeatin (found in corn kernels and coconut milk), isopentyl amino purine (2ip) are naturally occurring cytokinins in plants while kinetin, 6-benzyl purine (BAP) and thidiazuron are some commonly used synthetic cytokinins. A common practice observed during *in vitro* culturing of plants (referred to as plant tissue culture), is to add cytokinins to stimulate callus formation, which is an undifferentiated cluster of cells, to form shoots. Then the individual shoots are cut along with a portion of the callus grown in rooting medium resulting in the formation of genetically identical plantlets. Naturally, higher ratio of cytokinins to auxins results in shoot, compared to root, formation.

Gibberellins (GA)

Gibberellins are tetracyclic di-terpenoid compounds which include a large and diverse family of related compounds. More than 100 gibberellins have been isolated and denoted as GA1, GA2, GA3, and so forth. There are 126 known gibberellins, however, only four have been shown to have biological activity (i.e., GA1, GA3, GA4 and GA7). GA3 is the most common and biologically active gibberellin, most of the other non-active gibberellins are believed to be involved in the biosynthesis of the active gibberellins. GAs stimulate stem elongation and seed germination and promote shoot growth. GAs are produced in seeds, young shoots and developing fruits and flowers. Many forms of dormancy are broken by GAs facilitating growth of plants. Commercially, GA is used to enhance fruit size and the shape of fruits and to increase stem length and flowering in ornamentals. Treating barley seeds with GA is a common practice in the brewing industry to speed up the malting process. GA are used in seedless fruit production (e.g., grapes and oranges). GAs have applications in forestry to increase the cone production of pine and Douglas fir. In loblolly pine, GA improves the quality and size of cones. Lower concentrations of gibberellins can help root growth.

Abscisic acid

Until recently, abscisic acid (ABA) was believed to be the causative agent of abscission (leaf drop), however, it has been shown to play a critical role in modulating stress related responses. For instance, in tomato plants ABA is synthesized in roots when the plant is under water stress. As an adaptive response, ABA is transported to the leaves stimulating the closure of stomata. This mechanism helps the plant conserve water during periods of drought. ABA inhibits seed germination, acting opposite to the effect of GA. ABA promotes seed dormancy and helps seeds withstand desiccation by preventing premature germination under unfavorable conditions. In addition, ABA enhances the cold tolerance in plants by initiating dormancy in winter buds and, as the temperatures rises in spring, signaling the transition from a dormant phase to an active growing phase. Finally, ABA stops primary and secondary growth in vascular tissues allowing plants to conserve energy when conditions are not favorable for growth.

Ethylene

"One rotten apple can spoil the whole basket". This cliché has lingered in our language for decades, but is it true? Can one rotten apple really accelerate the rotting of other apples in the same basket?

Apples naturally produce and emit a gas called ethylene during the ripening stage. This is sensed by the other apples overstimulating a ripening response in the apples around the rotten one. Ethylene is effective at low concentrations as a plant growth regulator. In fact, ethylene gas is often used commercially to ripen harvested green fruits such as bananas, pears, mangoes, tomatoes and honeydew melons. To induce natural ethylene ripening, farmers sometimes wrap pears or peaches in paper, which keeps the gas from escaping. The relevance of ethylene goes beyond its influence on ripening fruits. Ethylene is also involved in aging and senescence of plants, including flowers. For floriculturists, managing ethylene exposure is vital to prolonging the life of cut flowers. It also impacts several reproductive related processes such as initiating flowering in pineapple and mango and promoting development of female flowers in cucumber buds. Ethylene also functions as a stress hormone and is produced in plants in response to a variety of biotic and abiotic stress conditions (discussed in Chapter 2).

Other, less studied, plant hormones include brassinosteroids, jasmonates, salicylic acid and polyamines. These substances can also elicit several physiological responses and are implicated as important regulatory molecules in plant defense. The functions of each of these hormones is summarized in Table 1.1.

While each hormone has its primary function, they often work synergistically or antagonistically to regulate various aspects of plant growth and development.

Table 1.1 Plant hormone types and functions

Plant hormone	Physiological action
Brassinosteroids-e.g., Brassinolide	Elongation promotion, stomata regulation, promotion of cell division, promotion of ethylene synthesis, anti-stress action, interact with other plant hormones
Jasmonates- e.g., Jasmonic acid and its derivatives	Elongation inhibition, wound response, leaf senescence induction, flowering inhibition, developmental changes during mechanical stress, photosynthesis inhibition, interact with other plant hormones
Salicylic acid	Plant stress resistance, protect photosynthesis machinery, maintain membrane integrity and regulation of ions including nutrients uptake
Polyamines- e.g., Spermidine	Cell division and differentiation, root elongation, DNA synthesis, programmed cell death, regulate ion homeostasis and transport, improve anti-oxidant defense systems, associated with abiotic stress tolerance, interact with other plant hormones

1.8 Plant secondary metabolites

Plants synthesize a large, structurally diverse range of specialized molecules that can contribute to their adaptation and successful survival in unfavorable conditions. These molecules are referred to as secondary metabolites and they include >200,000 compounds. Many of the secondary metabolites are accumulated in the vacuoles of plant cells.

The role of plant secondary metabolites (PSMs) is still not well understood; however, these molecules are mainly produced in response to various environmental stresses. Furthermore, they can promote beneficial plant-microbial interactions and often work together with plant hormones to coordinate responses to environmental stimuli. Noteworthy is the fact that beyond helping the plant itself, PSMs can sometimes provide immense medicinal benefits.

PSMs can be structurally divided into 3 major groups: alkaloids, terpenoids and phenolic compounds. In addition, smaller groups such as cyanogenic glucosides and glucosinolates are also produced in many plants. Terpenoids represent the largest group of PSMs and in some plants they are produced in high levels in specific tissues. For instance, in conifers the resin ducts store terpenoids and trap insects. Limonene is an example of a terpenoid that deters a leaf-cutting ant from destroying citrus plants Some terpenoids are present in floral fragrance and can act as attractants to pollinators or, conversely, repellants to predators. They are also known to contribute to the development of diverse range of flowers that we observe around us. Alkaloids play a crucial role in plant defense. Many alkaloids have been shown to be deleterious to herbivorous pests. Nicotine is an example of alkaloid that is toxic to beetles, caterpillars, and spider mites. It is produced in the roots of tobacco plants and is then transported to the leaves. Gramine, produced in the leaves of barley is toxic to aphids. Furthermore, some alkaloids have toxic effects against fungi and bacteria. Examples include liriodenine (produced by the tulip tree and is toxic to mushrooms) and chelerythrine (produced by celandine, a plant in the poppy family, and is toxic to bacteria). In addition to their insecticidal properties, many of the identified terpenoids and alkaloids have other biological activities (Table 1.2).

Table 1.2 Medicinal properties of some terpenes and alkaloids

Plant	Metabolite	Class	Functions
Coffee, tea	Caffeine	Terpene	Nervous stimulant
Palm tree	Arecoline	Terpene	Nervous stimulant
Opium poppy	Morphine	Terpene	Pain reliever
Maidenhair tree	Ginkgolides	Terpene	Cognitive
Ma Huang tree	Ephedrine	Alkaloid	Nervous stimulant
Pink periwinkle	Vinblastine	Alkaloid	Anti-cancer
Green snow drop	Galantamine	Alkaloid	Alzheimer's disease
Crossyne flava bulb	Pancratinine	Alkaloid	Parkinson's disease
Sweet wormwood	Artemisinin	Alkaloid	Anti-malaria
Chinese tree of joy	Camptothecin	Alkaloid	Ovarian cancer
Pacific yew	Taxol	Terpene	Anti-cancer
Cannabis	Friedelin	Terpene	Anti-inflammatory
Asiatic day flower	Homonojirimycin	Alkaloid	Anti-H_1N_1 virus
Decne tree	Camptothecin	Alkaloid	Anti-SARS-CoV2 virus

Phenolic compounds are important PSMs and exhibit a large variety of structures. They include flavonoids, tannins, and lignins which possess bioactive properties. Chalcones, flavones, flavanols, flavanones, anthocyanins, and isoflavones are major flavonoids and they perform a variety of roles (Fig. 1.25).

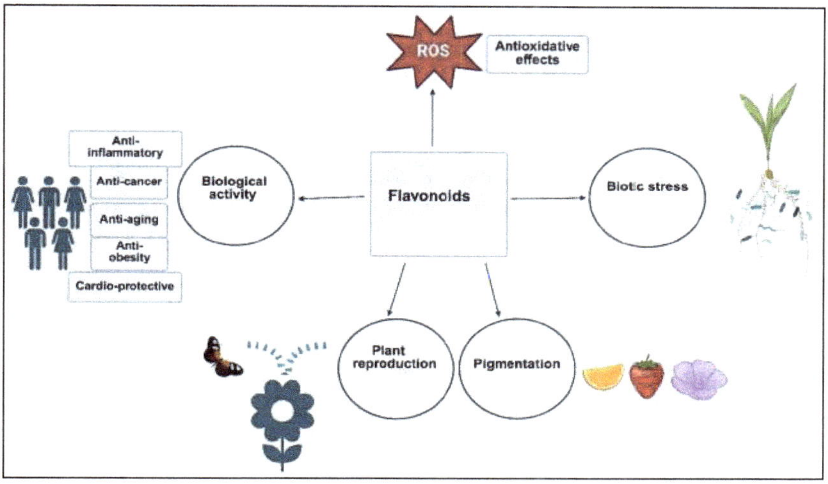

Fig. 1.25 Diverse functions of flavonoids

Lignin aids in plant resilience. It acts as a physical barrier to plant tissues, making them tough and less palatable or accessible to grazing animals. Tannins are bitter substances derived from gallic acid, the purpose of which is to deter herbivores. For instance, the gypsy moth is a pest that can devastate oak trees by eating its leaves until the tree dies. When a gypsy moth lands on a mature oak tree, the tree protects itself by sending tannins and enzymes to its leaves, which act as a deterrent to the moth. Remarkably, other oak trees in the area untouched by the moths will also send tannins to their leaves as a defensive strategy. Furthermore, the tannins bind with proteins in the digestive systems of the herbivores, making the plant material less digestible and less nutritious, which further discourages feeding. Tannins have numerous other biological activities and are increasingly being used in many pharmaceutical applications.

Cyanogenic glucosides and glucosinolates constitute a relatively minor group of PSMs and are induced in response to wounding or to feeding by grazing animals. Thus, damaged plants can be toxic to potential herbivores due to the activation of these compounds.

1.9 Further readings

Alberts, Bruce, ed. 2002. *Molecular Biology of the Cell*. 4th ed. New York: Garland Science.

Anjali, Sumit Kumar, Tulasi Korra, Rajneesh Thakur, R Arutselvan, Abhijeet Shankar Kashyap, Yasser Nehela, Victor Chaplygin, Tatiana Minkina, and Chetan Keswani. 2023. "Role of Plant Secondary Metabolites in Defence and Transcriptional Regulation in Response to Biotic Stress." *Plant Stress* 8 (June):100154. https://doi.org/10.1016/j.stress.2023.100154.

Augustine, J. J., L. R. Baker, and H. M. Sell. 1973. "Female Flower Induction on Androecious Cucumber, Cucumis Sativus L.1." *Journal of the American Society for Horticultural Science* 98 (2): 197–99. https://doi.org/10.21273/JASHS.98.2.197.

Bennett, Richard N., and Roger M. Wallsgrove. 1994. "Secondary Metabolites in Plant Defence Mechanisms." *New Phytologist* 127 (4): 617–33. https://doi.org/10.1111/j.1469-8137.1994.tb02968.x.

Bock, Ralph. 2014. "Genetic Engineering of the Chloroplast: Novel Tools and New Applications." *Current Opinion in Biotechnology* 26 (April):7–13. https://doi.org/10.1016/j.copbio.2013.06.004.

Castro-Camba, Ricardo, Conchi Sánchez, Nieves Vidal, and Jesús Mª Vielba. 2022. "Plant Development and Crop Yield: The Role of Gibberellins." *Plants* 11 (19): 2650. https://doi.org/10.3390/plants11192650.

Dewick, Paul M. 2009. *Medicinal Natural Products: A Biosynthetic Approach*. 1st ed. Wiley. https://doi.org/10.1002/9780470742761.

Ellis, R. John. 1979. "The Most Abundant Protein in the World." *Trends in Biochemical Sciences* 4 (11): 241–44. https://doi.org/10.1016/0968-0004(79)90212-3.

Faisal, Shah, Syed Lal Badshah, Bibi Kubra, Abdul-Hamid Emwas, and Mariusz Jaremko. 2023. "Alkaloids as Potential Antivirals. A Comprehensive Review." *Natural Products and Bioprospecting* 13 (1): 4. https://doi.org/10.1007/s13659-022-00366-9.

Ferreira, Maria-José U. 2022. "Alkaloids in Future Drug Discovery." *Molecules* 27 (4): 1347. https://doi.org/10.3390/molecules27041347.

Galuszka, Petr, Ivo Frébort, Marek Šebela, Pavel Sauer, Susanne Jacobsen, and Pavel Peč. 2001. "Cytokinin Oxidase or Dehydrogenase?: Mechanism of Cytokinin Degradation in Cereals." *European Journal of Biochemistry* 268 (2): 450–61. https://doi.org/10.1046/j.1432-1033.2001.01910.x.

Glick, Bernard R. 2005. "Modulation of Plant Ethylene Levels by the Bacterial Enzyme ACC Deaminase." *FEMS Microbiology Letters* 251 (1): 1–7. https://doi.org/10.1016/j.femsle.2005.07.030.

Guan, Ling, Yingjun Li, Kaihui Huang, and Zong-Ming Cheng. 2020. "Auxin Regulation and MdPIN Expression during Adventitious Root Initiation in Apple Cuttings." *Horticulture Research* 7 (1): 143. https://doi.org/10.1038/s41438-020-00364-3.

Jan, Rahmatullah, Sajjad Asaf, Muhammad Numan, Lubna, and Kyung-Min Kim. 2021. "Plant Secondary Metabolite Biosynthesis and Transcriptional Regulation in Response to Biotic and Abiotic Stress Conditions." *Agronomy* 11 (5): 968. https://doi.org/10.3390/agronomy11050968.

Kärkönen, Anna, and Sanna Koutaniemi. 2010. "Lignin Biosynthesis Studies in Plant Tissue Cultures." *Journal of Integrative Plant Biology* 52 (2): 176–85. https://doi.org/10.1111/j.1744-7909.2010.00913.x.

Khan, Zareen, David Roman, Trent Kintz, May delas Alas, Raymond Yap, and Sharon Doty. 2014. "Degradation, Phytoprotection and Phytoremediation of Phenanthrene by Endophyte *Pseudomonas Putida,* PD1." *Environmental Science & Technology* 48 (20): 12221–28. https://doi.org/10.1021/es503880t.

Khorasanizadeh, Sepideh. 2004. "The Nucleosome." *Cell* 116 (2): 259–72. https://doi.org/10.1016/S0092-8674(04)00044-3.

Kosová, Klára, Ilja Tom Prášil, Pavel Vítámvás, Petre Dobrev, Václav Motyka, Kristýna Floková, Ondřej Novák, et al. 2012a. "Complex Phytohormone Responses during the Cold Acclimation of Two Wheat Cultivars Differing in Cold Tolerance, Winter Samanta and Spring Sandra." *Journal of Plant Physiology* 169 (6): 567–76. https://doi.org/10.1016/j.jplph.2011.12.013.

Martins, Auxiliadora O., Adriano Nunes-Nesi, Wagner L. Araújo, and Alisdair R. Fernie. 2018. "To Bring Flowers or Do a Runner: Gibberellins Make the Decision." *Molecular Plant* 11 (1): 4–6. https://doi.org/10.1016/j.molp.2017.12.005.

Mierziak, Justyna, Kamil Kostyn, and Anna Kulma. 2014. "Flavonoids as Important Molecules of Plant Interactions with the Environment." *Molecules* 19 (10): 16240–65. https://doi.org/10.3390/molecules191016240.

Møller, Ian Max, Allan G. Rasmusson, and Olivier Van Aken. 2021. "Plant Mitochondria – Past, Present and Future." *The Plant Journal* 108 (4): 912–59. https://doi.org/10.1111/tpj.15495.

Narayanan, Zareen, and Bernard R. Glick. 2022. "Secondary Metabolites Produced by Plant Growth-Promoting Bacterial Endophytes." *Microorganisms* 10 (10): 2008. https://doi.org/10.3390/microorganisms10102008.

Normanly, J., J. P. Slovin, and J. D. Cohen. 1995. "Rethinking Auxin Biosynthesis and Metabolism." *Plant Physiology* 107 (2): 323–29. https://doi.org/10.1104/pp.107.2.323.

Patel, Ashish, Ravi Vanecha, Jay Patel, Divy Patel, Umang Shah, and Tushar Bambharoliya. 2022. "Development of Natural Bioactive Alkaloids: Anticancer Perspective." *Mini-Reviews in Medicinal Chemistry* 22 (2): 200–212. https://doi.org/10.2174/1389557521666210712111331.

Shitan, Nobukazu, and Kazufumi Yazaki. 2020. "Dynamism of Vacuoles toward Survival Strategy in Plants." *Biochimica et Biophysica Acta (BBA) - Biomembranes* 1862 (12): 183127. https://doi.org/10.1016/j.bbamem.2019.183127.

Sloan, Daniel B., Andrew J. Alverson, John P. Chuckalovcak, Martin Wu, David E. McCauley, Jeffrey D. Palmer, and Douglas R. Taylor. 2012. "Rapid Evolution of Enormous, Multichromosomal Genomes in Flowering Plant Mitochondria with Exceptionally High Mutation Rates." Edited by Michael William Gray. *PLoS Biology* 10 (1): e1001241. https://doi.org/10.1371/journal.pbio.1001241.

Sosnowski, Jacek, Milena Truba, and Viliana Vasileva. 2023. "The Impact of Auxin and Cytokinin on the Growth and Development of Selected Crops." *Agriculture* 13 (3): 724. https://doi.org/10.3390/agriculture13030724.

Tao, Hongyu, Ling Zuo, Huanli Xu, Cong Li, Gan Qiao, Mingyue Guo, and Xiukun Lin. 2020. "Alkaloids as Anticancer Agents: A Review of Chinese Patents in Recent 5 Years." *Recent Patents on Anti-Cancer Drug Discovery* 15 (1): 2–13. https://doi.org/10.2174/1574892815666200131120618.

Vanholme, Ruben, Rebecca Van Acker, and Wout Boerjan. 2010. "Potential of Arabidopsis Systems Biology to Advance the Biofuel Field." *Trends in Biotechnology* 28 (11): 543–47. https://doi.org/10.1016/j.tibtech.2010.07.008.

Wang, Shuting, Guodong Sun, Ying Luo, Wenjun Qian, Kai Fan, Zhaotang Ding, and Jianhui Hu. 2022. "Role of IAA and Primary Metabolites in Two Rounds of Adventitious Root Formation in Softwood Cuttings of Camellia Sinensis (L.)." *Agronomy* 12 (10): 2486. https://doi.org/10.3390/agronomy12102486.

Zandavar, Hamed, and Mojtaba Afshari Babazad. 2023. "Secondary Metabolites: Alkaloids and Flavonoids in Medicinal Plants." In *Herbs and Spices - New Advances*, edited by Eva Ivanišová. IntechOpen. https://doi.org/10.5772/intechopen.108030.

Chapter 2

Plant Biotic and Abiotic Stress

Plants growing in the natural environment are continually subjected to a wide range of both biotic and abiotic stresses. These stresses negatively impact different aspects of plant growth and development. Some of the biotic stresses include the presence of fungal and/or bacterial pathogens that are typically found in the soil but may also be airborne; insect predation attacking either plant roots, leaves or stems; and nematode predation which generally affects plant roots. Plants must also contend with a number of abiotic stresses including flooding, oxidative stress, drought, salt stress, heavy metal stress, light stress, temperature stress, and nutrient stress (Fig. 2.1). In the first instance, plants have developed a wide range of responses to these stresses so that, if the stress is not too severe or long lasting, plant growth and development may proceed more or less normally. More severe stresses may be addressed by creating transgenic plants (Chapter 6) that are resistant to specific stresses.

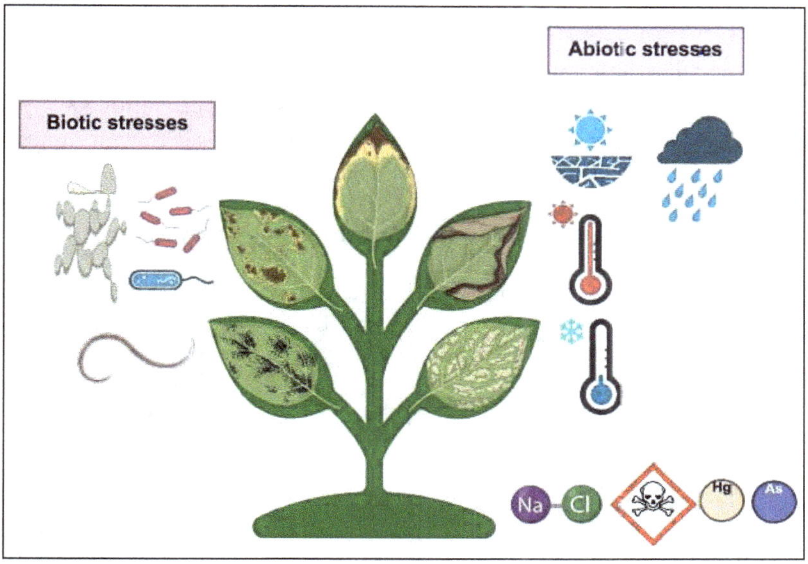

Fig. 2.1 Overview of biotic and abiotic stresses in plants

2.1 Fungal and bacterial pathogens

Plant pathogens (phytopathogens) significantly reduce the full potential of crop yields globally. Among these, bacteria and fungi are particularly impactful as phytopathogens due to their ability to readily replicate, spread quickly and adapt to diverse environmental conditions. Bacterial and fungal pathogens can reduce crop yields by as much as 30%, often leading to substantial economic losses and food security issues. Sometimes they can alter landscapes. For example, the iconic American elm tree was devoured by a fungus known as *Ceratocystic ulmi.* It was introduced into North America in the 1930's and subsequently wiped out about 77 million trees within a few decades altering the appearance of many landscapes, particularly in urban and suburban areas where American elm trees were commonly planted. Another prominent example is late blight, a fungal disease that attacks tomatoes and potatoes. This disease, caused by the fungal pathogen *Phytophthora infestans*, can rapidly destroy entire crops, leading to significant agricultural losses. In 2009, the "Great Tomato Blight" outbreak was seen in the northwestern United States, devastating tomato fields virtually overnight. Similarly, canker, a disease triggered by the bacterium *Xanthomonas axonopodis* pv. *citri* caused major outbreaks in citrus-growing regions, notably Florida (USA) while *Lasiodiplodia theobromae*, a fungus, killed millions of cacao (chocolate) trees in Central and South America.

Some other common plant diseases caused by fungal and bacterial species are given in Table 2.1. These diseases are manifest through various symptoms such as discoloration, wilting, defoliation, leaf spots, lesions leading to conditions such as powdery mildew, wilt, smut, blight, rot or crown gall (Fig. 2.2)

Table 2.1 Some diseases caused via bacterial and fungal phytopathogens

Pathogen type	Causative agent	Disease type	Distinguishing feature
Bacterium	*Burkholderia glumae*	Rot, blight	Rice pathogen
	Ralstonia solanacearum	Bacterial wilt	Sudden wilting of the whole plant; infected stems will ooze a slimy liquid
	Erwinia amylovora	Fire blight	Infected leaves and twigs display a characteristic 'shepherds crook'
	Dickeya dadantii	Soft rot	Produces plant tissue-macerating enzymes

	Agrobacterium tumefaciens	Crown gall	Modifies plant DNA
	Pseudomonas sp.	Holcus spot	Corn pathogen
Fungus	*Rhizoctonia solani*	Blight	Infects >200 plant species
	Botrytis cinerea	Grey mold	Turns bulbs and flowers spotty and brown
	Magnaporthe grisea	Rice ragged stunt and rice blast	Emerging rice pathogen
	Fusarium culmorum	Head blight, Foot and root rot	Contaminates plants and grains with mycotoxins; also, a post-harvest pathogen
	Puccinia sorghi	Rust	Common corn pathogen; spores are dispersed by wind infecting swaths of corn fields
	Hemileia vastatrix	Coffee leaf rust	Coffee pathogen
	Fusarium oxysporum	Fusarium wilt	Major recurring pathogen of cotton
	Plasmopara viticola	Downy mildew	Appears as pale yellow fuzzy, feathery mold on leaves

Fig. 2.2 Some common symptoms due to fungal and bacterial plant pathogens (a) Smut: a disease common in grains, turns plants into a black powdery mass, (b) Rust: causes brown spots on leaves, (c) Wilt: wilting of a tomato plant, (d) Rot: discoloration of a conifer, (e) Canker: reddish structures on the bark of a tree, (f) Powdery mildew: leaf covered with white, powdery fungal spores, and (g) Crown gall: galled tissue of a mango tree.(photo credit: https://www.missouribotanicalgarden.org/)

When attacked by a potential pathogen, plants typically mount a strong defense response both locally and throughout the plant. Local resistance occurs in the form of necrotic lesion(s) at the site of infection. This localized response, termed the hypersensitive response, can be triggered by a wide variety of pathogens and occurs within a few hours of infection consequently forming a necrotic lesion or host cell death. Systemic acquired resistance (SAR) is developed after the initial attack of the pathogen and provides resistance to the pathogen in parts of the plant distant from the site of infection. SAR also induces resistance against other (generally similar) pathogens after the initial pathogen attack.

Fungal and bacterial plant pathogens appear to be recognized by pattern recognition receptors (PRRs) on the plant cell plasma membrane, which identify pathogen-associated molecular patterns (PAMPs). These PAMPs are conserved molecular signatures found in a wide range of pathogens. When PRRs detect these PAMPs, they trigger an "immune response" in the plant, leading to the activation of various defense mechanisms (Fig. 2.3). The second level of pathogen recognition involves plant resistance (R) genes, which identify specific receptors from a pathogen. R genes operate via gene-for-gene interaction between the pathogen and its host plant, unlike the broad-spectrum recognition of PAMPs by PRRs. This gene-for-gene hypothesis is broadly applicable to many plant-pathogen interactions, subsequently activating hypersensitive defense responses and triggering localized cell necrosis. Additionally, this recognition can activate SAR, enhancing the overall ability of the plant to combat pathogen attack and prevent widespread infection.

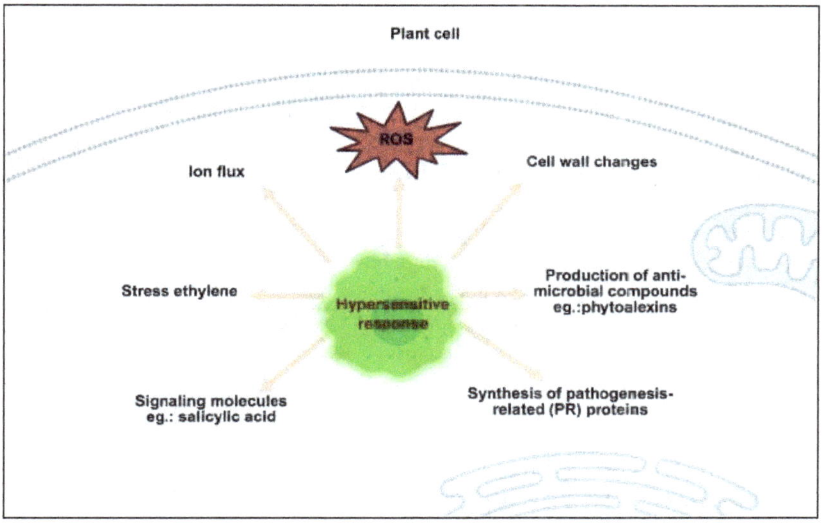

Fig. 2.3 An overview of a plant cell undergoing a hypersensitive response releasing signals for local and systemic induction of defense responses

Plant cells are in constant chemical communication with one another and with their environment. They recognize and respond to external stimuli using a system of chemical messengers that receive and transmit the stimuli initiating a response. The response may be stimulatory or inhibitory, for example following pathogen infection plants synthesize 'stress ethylene' which triggers a stress or senescence response that inhibits the spread of the pathogen within the plant. At first, a small amount of stress ethylene is produced rapidly to initiate the plant's synthesis of defensive proteins. However, the accumulation of 1-aminocyclopropane-1-carboxylate (ACC), the immediate precursor to ethylene leads to the buildup of unwanted stress ethylene. At this point, ethylene becomes inhibitory to plant growth and can exacerbate some of the damage caused by the pathogen.

One of the key signaling molecules, which helps in activating SAR in plants is salicylic acid. Plants convert a conjugated storage form of salicylic acid (2-o-β-D glucosylsalicylic acid or SAG), to salicylic acid, which quickly activates SAR. Salicylic acid is produced in sensing tissues and moves through the phloem sap to distal tissues to activate SAR elsewhere in the plant. Additionally, a group of pathogenesis-related (PR) proteins with anti-pathogen activities such as antibiotics and fungal cell wall degrading enzymes are induced. Some examples of PR proteins with their distinctive roles are summarized in Table 2.2.

Table 2.2 PR proteins and their mechanism of action

Protein name	Type	Mechanism of Action	Target Fungi	Examples
PR1	PR protein	Inhibits fungal growth through membrane disruption	Various fungi	Tobacco, Arabidopsis
β-1,3-glucanase (PR2)	PR protein	Breaks down β-1,3-glucans in fungal cell walls	Fungi with glucans	Barley, wheat
Chitinase (PR3)	PR protein	Degrades chitin in fungal cell walls	Chitin containing fungi	Tomato, rice
Thaumatin-like proteins (PR5)	PR protein	Disrupts fungal membranes	Broad spectrum	Apple, grapevine
Defensins (PR12)	Peptide	Disrupts fungal cell membranes	Broad spectrum	Pea, radish
Thionins (PR13)	Peptide	Disrupts fungal cell membranes	Some fungi	Barley, wheat
Lipid Transfer Proteins (LTPs) (PR14)	Protein	Disrupts fungal membranes; inhibits spore germination	Broad spectrum	Corn, soybean
Protease Inhibitors	Protein	Inhibits fungal proteases, affecting growth	Broad spectrum	Potato, tomato
Lectins	Protein	Binds to carbohydrates on fungal cells, disrupting growth	Some fungi	Beans, pea
Pectate Lyase	Enzyme	Degrades pectin in fungal cell walls, disrupting structure	Pectin containing fungi	Citrus, tomato
Major latex like proteins (MLP's) (PR10)	Protein	Can disrupt fungal cell membranes	Various fungi	Opium poppy

The other initial response of pathogen infection on plants includes accumulation of antimicrobial compounds such as phytoalexins, which are low molecular weight secondary plant metabolites. A number of phytoalexins have been found to be inhibitory to the growth of phytopathogenic fungi and bacteria. High sugar concentrations can also limit pathogen growth in infected tissues. Plants can accumulate sugars, thereby enhancing resistance to pathogens.

2.2 Insect and nematode predation

A farmer's success hinges on crop yield, which can be impacted by several factors. Plant diseases caused by pests can cause huge economic losses in addition to ruining people's livelihoods. For instance, the boll weevil (the beetle *Anthonomus grandis*) which infected cotton (a primary cash crop) in the American South in the 1920's, caused millions of dollars of damage and also led to significant hardships to many farmers and their families who depended on this crop.

Currently, insect and nematode infestations are responsible for around a 14% decrease in crop production each year, accounting for approximately 30-50% of the total loss plants incurred worldwide. Currently, synthetic insecticides are largely used to suppress the damage caused by these pests. These substances can be derived from chemicals or biological material. Some common insecticides, belonging to groups including organochlorines, organophosphates and carbamates, can affect a wide variety of insect species. One of these, DDT, an organochlorine insecticide, which was invented in the 1940's proved to be highly effective in killing numerous species of insect pests. However, there were many challenges associated with its use, among which bird toxicity and persistence in the environment were significant concerns. Organophosphates act on the insect's nervous system (cholinesterase inhibition) and were thought to be good alternatives to DDT but they can adversely affect non-target organisms. For instance, chlorpyrifos is one of the most widely used organophosphate insecticides on crops such as cotton, corn, fruit and nut trees; however, its exposure causes various nervous disorders in humans. It is also highly toxic to amphibians, fish, beetles and birds. In addition, plants grown in soil that is contaminated with insecticides may transfer the chemical to those who consume the plants leading to potential health risks.

All plants possess a certain degree of resistance to insects, developed through a combination of physical, chemical, and biological defense mechanisms. Physical defenses such as thick cell walls, trichomes (hair-like structures), and cuticles deter insect feeding and penetration. Production of plant-protective secondary metabolites (as discussed earlier) in a wide range of plants are a part of plants' natural defense system, and can be toxic, repellent or anti-nutritional to insects, effectively reducing herbivory. Some plants inherently produce proteins such as protease and α-amylase inhibitors that disrupt the digestive systems of the feeding insects by preventing the breakdown of plant proteins or plant carbohydrates thereby limiting nutrient absorption. Some plant proteins are directly toxic to insects, for instance lectins which are carbohydrate-binding proteins are found in the seeds and storage tissue of many plant species.

Nematodes are tiny parasites that damage food and fiber crops worldwide and cause billions of dollars of losses annually. Some nematode pathogens are ectoparasites, meaning they live outside their plant host. Endoparasites are those species that reside inside roots and cause cellular necrosis. The root-knot nematode and the cyst nematode are the major endoparasites causing the most economic damage. For instance, the soybean cyst nematode (*Heterodera glycines*) is a significant pathogen affecting soybean crops and the potato cyst nematodes (*Globodera pallida* and *G. rostochiensis*) cause substantial yield losses in potato production throughout the world. Root-knot nematodes (*Meloidogyne* spp.) affect a wide range of plants and are known for causing characteristic galls or "knots" on the roots of infected plants (Fig. 2.4). *Meloidogyne graminicola*, a common root-knot nematode species, is one of the major rice pathogens and can cause yield losses from 17% to 32%. Several key steps are involved from the primary infection stage to the final damage caused by endoparasitic nematodes. These nematodes develop a close association with the host plant, causing morphological changes in the host root cells that turn into feeding cells for the nematode. These modified root cells serve as a permanent nutrient source for the nematode. Any damage to the feeding cell can disrupt the lifecycle of the parasite. Some plants respond by accumulating callose, a polysaccharide, between the plasma membrane and the cell wall at the site of nematode infection. This forms a physical barrier that can prevent the spreading of the nematode. Plants also increase the deposition of extensins, a type of glycoprotein in the cell wall, to prevent pathogens from gaining access inside the cell. Nematode R-genes have been characterized in plants including tomato (e.g., the *Mi*-R gene confers resistance to both aphids and nematodes) and some cereal crops.

Fig 2.4 Plant infected with root-knot nematodes showing galls. (Photo credit: https://www.missouribotanicalgarden.org/)

2.3 Flooding and oxidative stress

When flooding occurs, the soil becomes saturated with water, displacing air from the soil pores. This lack of air (anaerobic conditions) reduces the availability of oxygen in the root zone causing a condition known as hypoxia or oxygen-deficient stress. Plants can survive to some extent during hypoxia by altering their metabolism to glycolysis and fermentation. However, prolonged exposure to hypoxia results in plant injury due to a build-up of toxins.

Hypoxia is a serious impeding factor for plant growth. Flooding stress substantially alters plant physiology and metabolism. Flooding can cause an outburst of reactive oxygen species or ROS in plant cells. ROS, include superoxide anion ($O_2^{\bullet-}$), hydroxyl radicle (•OH), hydrogen peroxide (H_2O_2) and singlet oxygen($^1O_2$) which are produced in many cellular compartments are extremely harmful to plant cells at high concentration (Fig. 2.5). Normally, cells need ROS to communicate with each other and respond to stimuli, but ROS can be detrimental if their levels are not regulated. When excess ROS is formed under stressed condition, the cell subsequently experiences 'oxidative stress' which can damage lipids, proteins and DNA. Membrane lipid peroxidation occurs when ROS reacts with unsaturated fatty acids leading to leakage of cellular components, desiccation and ultimately cell death.

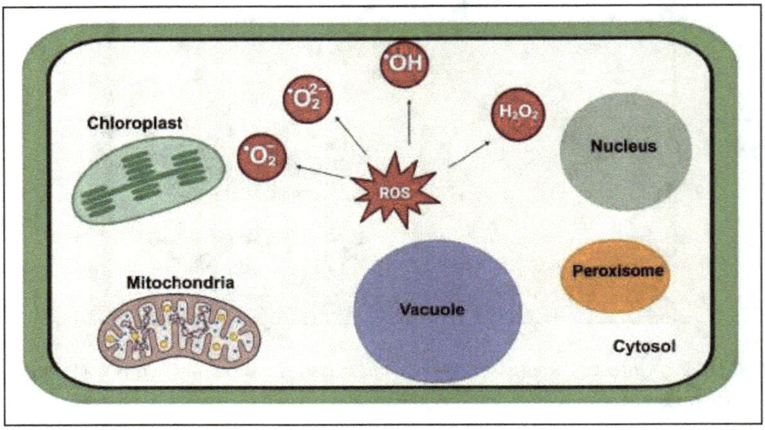

Fig. 2.5 Sites of reactive oxygen species (ROS) production in plant cells

In plants, the biosynthesis of ACC and ethylene follows a relatively straightforward pathway, starting with the compound S-adenosylmethionine (SAM) being converted to ACC by the enzyme ACC synthase followed by ACC being converted to ethylene by the enzyme ACC oxidase (Fig. 2.6). Excess ACC may be converted into the storage compounds 1-(malonyl)-ACC (MACC) or 1-(glutamyl)-ACC (GACC).

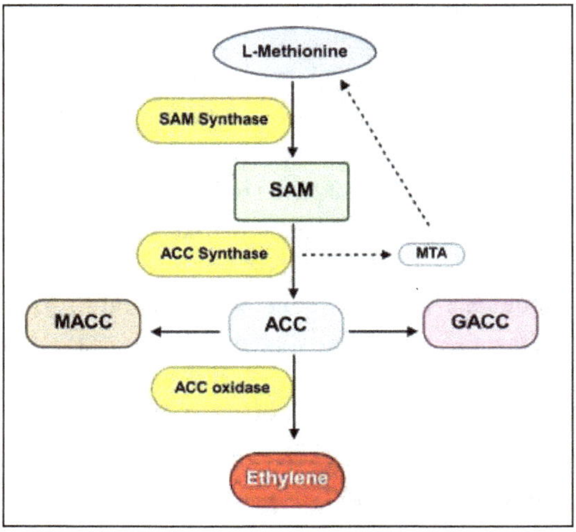

Fig. 2.6 Overview of the biosynthesis of ethylene in plants. Abbreviations: SAM S-adenosyl methionine; MTA 5′-methylthioadenosine; ACC 1-aminocylopropane-1-carboxylate; MACC 1-(malonyl)-ACC; GACC 1-(glutamyl)-ACC

Under oxygen deprived (as well as other abiotic stress) conditions, plants upregulate the production of the enzyme ACC-synthase, resulting in an increased accumulation of ACC in root cells. The lack of oxygen in roots prevents the conversion of ACC to ethylene so that ACC gets transported via the xylem to shoots where oxygen is present and the ACC is oxidized to ethylene, consequently inducing epinasty (extensive wilting) (Fig. 2.7), thereby, negatively impacting photosynthesis and overall plant growth. Furthermore, production of stress ethylene induced by flooding of the root system prevents root elongation as a consequence of ethylene interference with auxin action. Thus, the plants' ability to explore deeper soil layers for nutrient exploration is negatively affected.

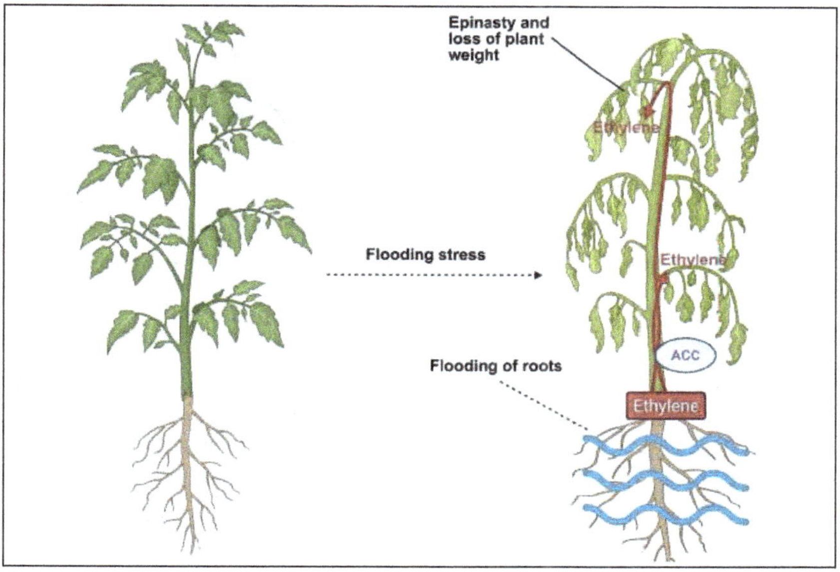

Fig. 2.7 Diagrammatic representation of how flooded roots (shown on the plant on the right) cause epinasty in plants. The control (non-flooded) plant is shown on the left

Some plants rely on morphological adaptations to tolerate flooding (Fig. 2.8). For example, rice plants can develop extensive aerenchyma in their roots, enabling them to transport oxygen from above-water parts of the plant to submerged roots. This adaptation allows rice to thrive in paddy fields, which are often flooded. Mangroves grow in coastal, waterlogged environments. They develop specialized adventitious roots called pneumatophores, which stick out of the water and allow the plant to take in oxygen directly from the atmosphere.

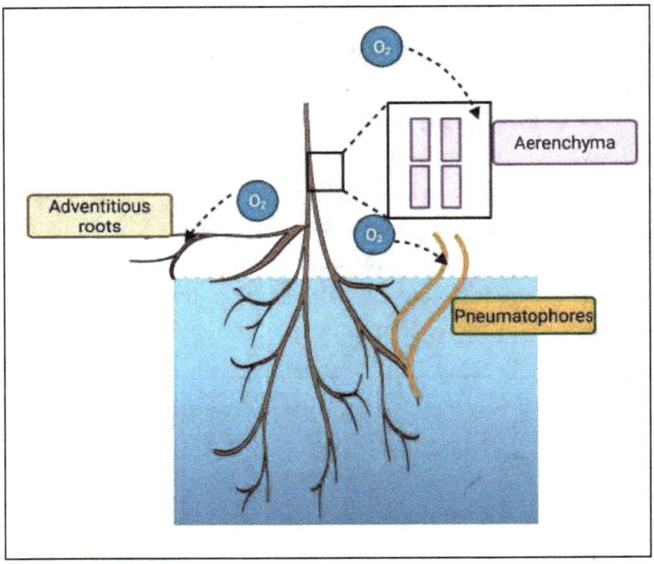

Fig. 2.8 An illustration of morphological adaptation in submerged plants. Formation of adventitious roots, pneumatophores and aerenchyma facilitate transport of oxygen to roots under hypoxic conditions

Additionally, enzymatic anti-oxidants such as superoxide dismutase (SOD), guaiacol peroxidase (GPX), ascorbate peroxidase (APX), catalase (CAT), glutathione reductase (GR) as well as non-enzymatic anti-oxidants such as ascorbate, α-tocopherol, glutathione (GSH), carotenoids, alkaloids, tocopherols, proline, and phenolic compounds (flavonoids, tannins, and lignins) that act as the scavengers of free radicals provide the plant with a certain degree of tolerance to oxidative stress.

2.4 Drought and salt stress

Drought stress has a profound impact on crop production, so much so that the resulting crop losses are greater than the total effects of all other plant stresses combined. This is a challenging issue not just for growers but also for researchers, policy makers, and communities that depend on agriculture for their livelihoods. Consequently, the food supply is affected, and many nations are facing growing food insecurity. According to the Food and Agricultural Organization of the United Nations (FAO), drought is a major factor contributing to reduced agricultural productivity (especially in the developing world), which in turn exacerbates food insecurity.

Plants cope with water scarcity in many ways. For example, pine needles are adapted to a xeric (dry) environment. Needles have a thick cuticle which protects the mesophyll (inner tissue) from drying out. Some plants adapt via drought escape (e.g., early flowering) or by suppressing drought-induced senescence and enabling these plants to resume normal growth once environmental conditions improve. Development of extensive root systems allows plants to access water from deeper soil layers, making them less dependent on surface moisture.

Drought stress can activate phytohormone biosynthesis, increasing the concentration of certain plant hormones. For example, abscisic acid (ABA) plays a critical role in drought tolerance and acts as a signaling molecule inducing stomatal closure (Fig. 2.9). By regulating the opening and closing of stomata, plants can significantly reduce transpiration, thereby minimizing water loss. Plant cells must maintain water balance and turgor pressure during periods of drought. This is also accomplished via the production of low molecular weight solutes (sugars) and osmolytes (such as trehalose, glycine betaine, proline, and polyamines). Some of these molecules accumulate at high levels and help to maintain cellular osmotic balance and protect cellular structures from dehydration. In particular, trehalose has the ability to form a gel-like matrix during desiccation, helping to maintain membrane integrity thereby keeping the cells function properly. Moreover, the synthesis of many of these osmolytes are also regulated by ABA. By activating the stress-responsive genes, many of which are involved in osmolyte metabolism, water retention, and ROS scavenging, ABA can counter many of the adverse effects associated with drought stress. Additionally, ABA induces the shedding of leaves, a process known as leaf abscission, which further helps to reduce water loss during periods of water scarcity. Hence, the co-operative regulation of ABA synthesis and osmotic adjustment mechanisms plays a key role in plant development during drought stress.

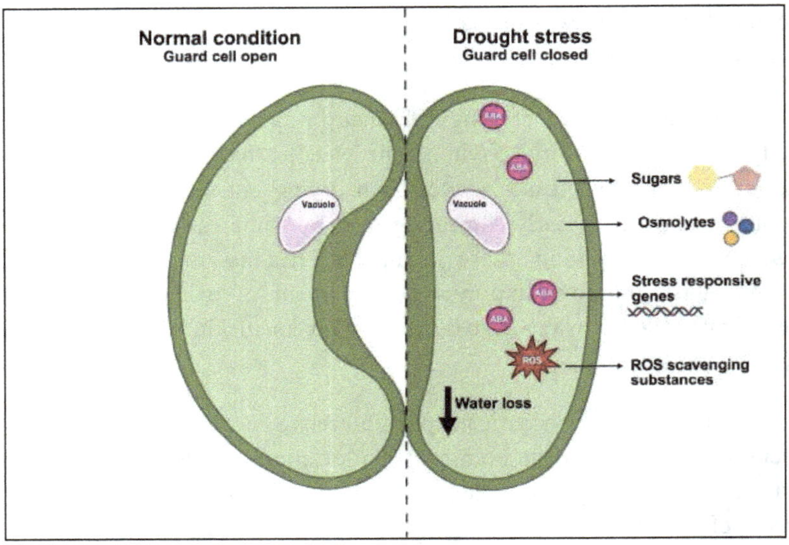

Fig. 2.9 An overview of possible drought tolerance mechanisms in plants

Salinity is increasingly becoming a major limiting factor to agriculture and food security, as approximately 45 million hectares of irrigated land are experiencing saline stress worldwide. It is estimated that more than 50% of global arable land could be negatively salt affected by the year 2050 with the problem expected to worsen over time due to poor irrigation practices, frequent salting of snowy roads during winter time, industrial expansion and various agricultural malpractices. Different plant species exhibit varying levels of salt tolerance, for instance wheat and rice are less tolerant than barley. Tomatoes and corn are sensitive to salt, and high salinity can limit their growth and yield. Cotton, a fiber crop, grown in arid and semi-arid environments globally is somewhat salt-tolerant; however, high levels of salt can limit its growth, fiber quality, and yields. On the other hand, *Frankenia pulverulenta* L. (Fam. Frankeniaceae) a small annual plant found in coastal habitats is halophytic, which means it can tolerate high levels of salt in this case due to the presence of salt glands in its leaves that exclude excess levels of salt, visible as salt crystals on the leaf surface (Fig. 2.10).

Fig. 2.10 Salinity stress tolerance in halophytes (a) Formation of salt crystals on the leaves of *F. pulverulenta* (b) a zoomed-in image of the leaf surface showing salt crystal (adapted from Bueno *et al.* 2021)

High soil salinity can impose two stress factors on plants. First, an osmotic stress, which is caused by increase in osmotic pressure in the soil due to reduced water availability, second, an ionic stress resulting from a solute imbalance, increasing the concentration of toxic Na^+ and Cl^- in the plant cytoplasm. Furthermore, plants have to cope with water loss caused by osmotic stress, and the accumulation of salt prevents nutrients from reaching the roots. Therefore, both plant growth and yield are negatively affected. Salt tolerance mechanisms in plants involves the following processes: (i) restriction of uptake and transport of toxic Na^+ and Cl^- ions from soil (ii) exclusion of Na^+ and Cl^- ions from the cytoplasm, and (iii) concentration of salt in the plant's vacuoles (large, storage compartments in plant cells). This is typically carried out by transmembrane transport proteins that exclude Na^+ from the cytoplasm both at the root and shoot level. The ability to sequester salt in the roots has been associated with reduced salt accumulation in shoots, contributing to salt tolerance. On the other hand, excess salt can induce cell senescence, thereby impairing the cell's ability to maintain vacuolar sequestration of sodium.

In many plant species, apoplastic barriers serve to prevent transport of ions (e.g., sodium) into the root cortex. In *Oryza sativa* (rice), roots can strengthen their apoplastic barrier by depositing materials such as lignin, in response to salt stress, which, in turn, contributes to a reduction in uptake and accumulation of sodium in the shoots. Recently, this has been shown as a predominant mechanism in woody nut crops such as almonds. Similar to drought stress, osmotic adjustments can assist plants in maintaining cell turgor under salt stress. When plants experience salt stress, the level of osmolytes such as trehalose increase to balance the osmotic pressure, thereby preserving cell turgor and preventing dehydration. This adjustment helps the plant maintain its structural integrity and continue physiological processes despite the presence of high salt concentrations.

Apart from the above-mentioned strategies, redistribution of sodium from young tissues to older 'sinks' or sodium recycling back to the soil are suggested as other processes that can enhance the ability of plants to combat salt stress (Fig. 2.11).

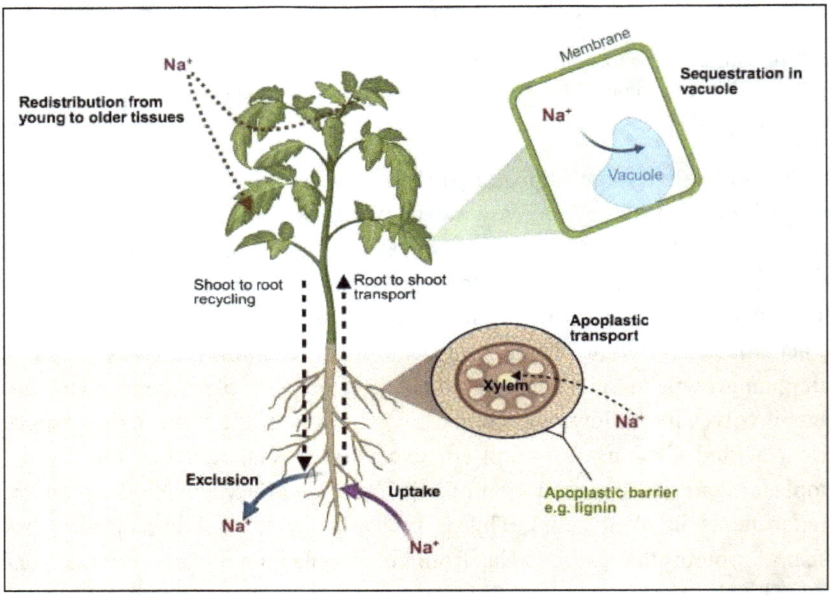

Fig. 2.11 Representation of mechanisms contributing to salt tolerance in plants

2.5 Heavy metal stress

Heavy metal pollution is a major problem in both developed and developing countries where industrial wastes and pollutants are directly dumped into rivers or left open on soil surfaces. Accumulation of heavy metals through the soil and irrigated water can significantly hamper normal plant function. Among numerous heavy metals, four heavy metals arsenic (As), lead (Pb), cadmium (Cd), and mercury (Hg) are considered as the most toxic metals interrupting many biological functions in plants. Some examples of toxic symptoms induced by heavy metals are listed in Table 2.3.

Table 2.3 Some examples of phytotoxic effects due to heavy metal exposure

Metal	Symptoms
Arsenic -As(III) and As(V)	Cellular damage and oxidative stress (due to ROS production) resulting in leaf burn and dead tissue
Lead (Pb)	Lipid peroxidation of membranes, membrane leakage, change in lipid composition; DNA damage impacting chlorophyll synthesis, cell division, seed germination
Cadmium (Cd)	Impacts enzyme functions involved in photosynthesis resulting in yellowing of leaves, stunted growth and leaf epinasty
Mercury (Hg)	Binds to water channel proteins leading to stomatal closure and interrupting water flow through cells; oxidative stress causing root damage and purpling of leaves
Copper (Cu), zinc (Zn), uranium (U), manganese (Mn), aluminum (Al)- ***(In excess)***	Inhibiting root elongation, decreasing photosynthesis; altering the amounts of essential micronutrients in cells disrupting normal physiological processes

Some plants are metal hyperaccumulators, meaning they can absorb large amounts of heavy metals, mostly as a strategy to defend against herbivores or pathogens. And they are able to do so due to metal transporters. Due to the similarity in chemical properties between essential and non-essential metals, these transporters can inadvertently transport toxic metals as well as essential metals. The toxic metals are sequestered in cell vacuoles where they can bind to chelating agents and remain in the plant throughout its life, reducing the toxicity to the plant itself. These plants, therefore, can be useful in the removal of toxic metals, and have been used in clean-up of heavy-metal polluted sites. This technology is referred to as phytoextraction (Fig. 2.12) and is commonly used as to restore metal-impacted sites About 450 flowering plant species are known to be hyperaccumulators, and, as more plants are tested, this number is growing. Metals can commonly be found in serpentine soils (rich in Ni, Co, Cr, and Mn), calamine soils (rich in Pb, Zn, and Cd), copper-bearing soils and in the island of New Caledonia, where a remarkable number of nickel hyperaccumulator plants are found.

There are other components that play a vital role in the detoxification process of heavy metals. For instance, the cell wall being the first line of defense has polysaccharides that form complexes with heavy metals thereby immobilizing them and preventing their entry into heavy metal-sensitive areas within the plant cell preventing subsequent interference with the normal functioning of the cell. Besides, specific amino acids (e.g., proline and histidine); hormones including salicylic acid and jasmonic acid also aid in detoxification. At the molecular level, many of the tolerance mechanisms are controlled by hundreds of genes involved in stress signaling pathways, processes governing various physiological functions, transcription factors that help in regulating anti-defense systems and hormone biosynthesis and in the synthesis of a host of other proteins. Together, these factors can strengthen a plant's resistance against heavy metal stress.

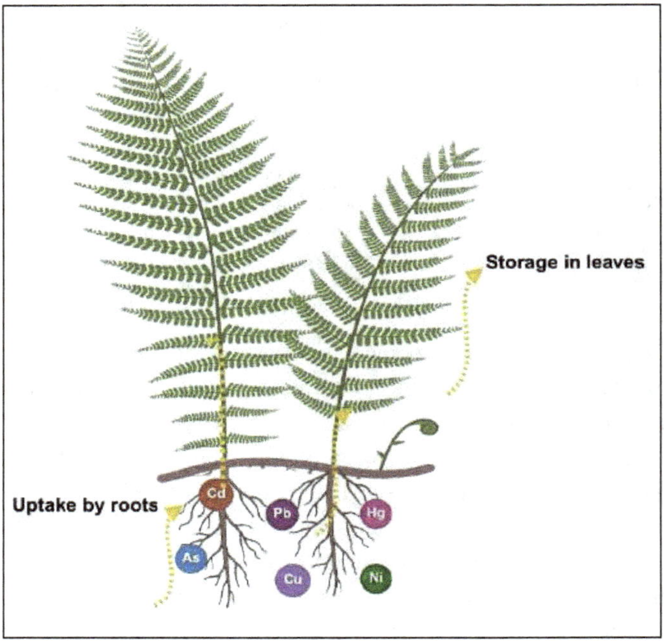

Fig. 2.12 Schematic diagram of the phytoextraction process. Plants can absorb heavy metals like arsenic, copper, nickel, cadmium, and lead from the soil through their roots. A portion of these metals are then translocated to the above-ground parts of the plant, including leaves and stems

2.6 Light stress and temperature stress

If a plant has plenty of warm air but no sunlight, it will not thrive. Light drives photosynthesis, which involves photosynthesis pigments, the Calvin cycle, and light-driven electron transport (photosystem I & II). Both the quality (color or wavelength of light) and quantity of light influence plant growth. Plants prefer blue light (425-490nm) and red light (610-700nm). Generally, the more sunlight a plant receives, the greater its capacity for producing food via photosynthesis, however, when the light intensity exceeds the optimal range for photosynthesis, it causes abiotic stress and physiological damage (Fig. 2.13). Under high light conditions, chlorophyll molecules absorb excess photons, leading to high-energy excited states (too much heat) potentially leading to cell damage. High light intensity also results in oxidative stress due to ROS accumulation in chloroplasts which damages the photosynthetic machinery and other cellular components.

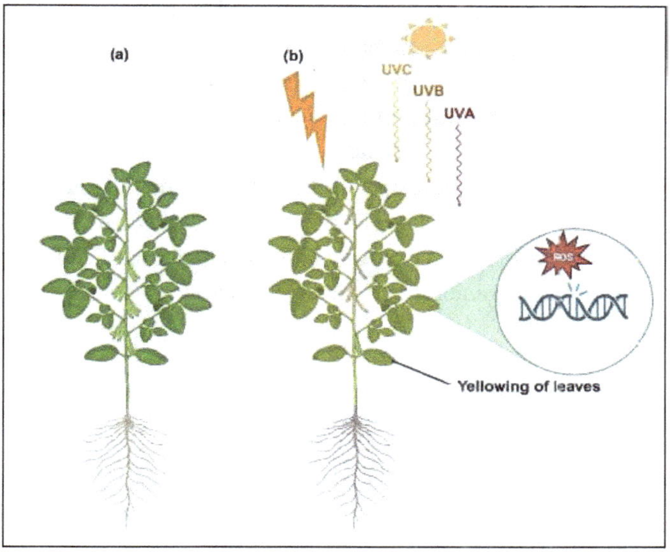

Fig. 2.13 Schematic illustration indicating (a) plant growth under normal conditions (b) plant growth under light stress

To avoid absorption of excessive sunlight, plants move their chloroplasts to the cell walls parallel to the direction of the light. This positioning minimizes the surface area exposed to the light, thereby reducing light absorption and protecting the chloroplasts from photodamage. Conversely, under low light conditions (mild stress) chloroplasts distribute themselves along the cell walls perpendicular to the incoming light to maximize light absorption for photosynthesis. In addition, to manage excess light a phenomenon called non-photochemical quenching or NPQ is activated in plants, where the excited states of chlorophyll are quenched rapidly. NPQ involves several components, some of which are triggered due to the acidification of the thylakoid lumen during high light conditions.

Plants growing under direct sunlight can experience stress by ultraviolet (UV) light which can damage the DNA. Certain UV-absorbing compounds such as flavonoids and phenolics produced in the vacuoles of the leaf epidermis can protect plants to some extent against this damage.

Temperature stress classified as high temperature and low temperature stress can significantly affect plant traits due to the following factors:

- Temperature plays a critical role in plant metabolism with many plant processes being tightly regulated by temperature.
- Low temperatures can lead to reduced enzyme activity, rigidification of membranes, destabilization of protein complexes, stabilization of RNA secondary structures, accumulation of ROS, impairment of photosynthesis, and leakage across membranes.
- High temperatures cause proteins to unfold and aggregate resulting in metabolic imbalances, membranes to become more fluid (increase in Ca^{+2} influx), and changes in the organization of cellular structures, including organelles and the cytoskeleton.

Survival in temperature extremes can be achieved in a variety of ways (discussed in the next section), however if the conditions become severe, it may lead to cellular damage or cell death within minutes.

2.7 Heat and cold stress

Temperatures greater than 75°F (considered optimum threshold) are referred to as causing heat stress in plants. While tropical plants are generally well-adapted to temperatures between 75°F and 95°F, temperatures above 95°F can push tropical plants past their tolerance limits, especially if accompanied by drought stress. The susceptibility to high temperatures varies with the stage of plant development. An increase in the duration of heat stress may cause irreversible changes to plant cells, leading to various signs of heat stress such as wilting and leaf damage. Heat stress particularly lowers the photosynthetic capacity of plants. Photochemical reactions, thylakoid membrane fluidity, chlorophyll biosynthesis, and chloroplast structure are all sensitive to heat. The adverse impact of heat stress on plants is due to the rapid outburst of ROS that can cause changes in ion homeostasis and enzyme inactivation. Rubisco, the key enzyme responsible for carbon fixation in the process of photosynthesis is also sensitive to temperature. Elevated temperatures can lead to the denaturation of Rubisco, reducing its enzymatic activity and efficiency. As heat stress intensifies, the enzyme's ability to catalyze the conversion of carbon dioxide into organic compounds is impaired, which directly impacts the rate of photosynthesis leads to imbalances in metabolic pathways, which can result in reduced nutrient transport and ultimately, plant tissue damage or death.

Similar to drought stress, a plant's response to heat could result in accelerated stomatal closure and reduced water loss. Compounds such as sugars, sugar alcohols, proline, glycine betaine, and tertiary quaternary ammonium compounds which typically participate in abiotic stress responses also contribute to heat-stress tolerance. In response to heat stress, plants can regulate through molecular chaperones and the rapid synthesis of heat shock proteins. These assist in stabilizing proteins when temperatures rise, ensuring that they don't become misshapen or lose their function. This is especially important during heat stress, as properly folded proteins are crucial for the cell's health and function.

Plants that are adapted to the cold temperatures of their growing environment are called 'hardy'. Winter injury in non-hardy plants can occur due to exposure to cold stress which may be of two types, chilling stress (0-15°C) and freezing stress (below 0°C). Minor chilling stress is normally reversible, but frost can destroy an entire crop. Freezing injury in plants is mainly caused by dehydration of plant cells as a consequence of ice buildup, so that drought and freezing tolerance may share some common protective mechanisms. When temperatures drop and freezing begins, ice forms outside of the cells. As ice continues to grow, more and more cells lose water and become dehydrated (Fig. 2.14). If the cells stay dehydrated for too long, the plasma membrane (which is the primary site of freezing injury) becomes permeable, and materials leak in and out, resulting in cell death. As more cells die, the likelihood of the plant recovering decreases.

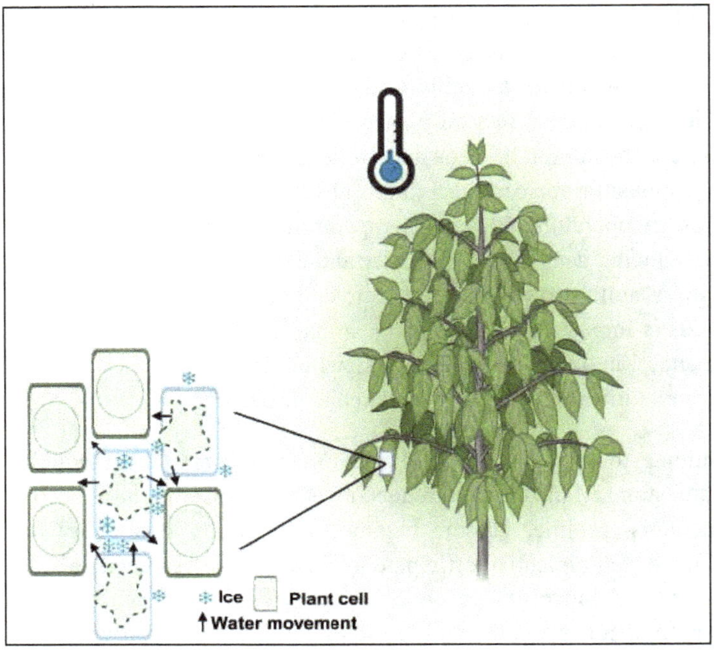

Fig. 2.14 Schematic representation of inter and intracellular freezing of plant cells

Some plants are better than others at tolerating frost stress. For instance, olive trees accumulate significant amounts of mannitol, a sugar alcohol that helps lower the freezing point or reduce the size of ice crystals which may form. Aquaporins, which are protein channels that facilitate movement of water molecules in and out of the membrane (Fig. 2.15), can effectively prevent ice formation inside cells and can also help to maintain membrane integrity. Low temperatures can cause membranes to become too rigid, but unsaturated fatty acids enhance membrane fluidity, preventing them from becoming stiff and brittle. Together, all these factors can help plant cells to better withstand freezing temperatures. Moreover, a variety of cold-regulated (COR) genes involved in hormonal biosynthesis, carbohydrate metabolism, membrane lipids, and polysaccharides as well as antifreeze proteins also confer cold tolerance in some plants.

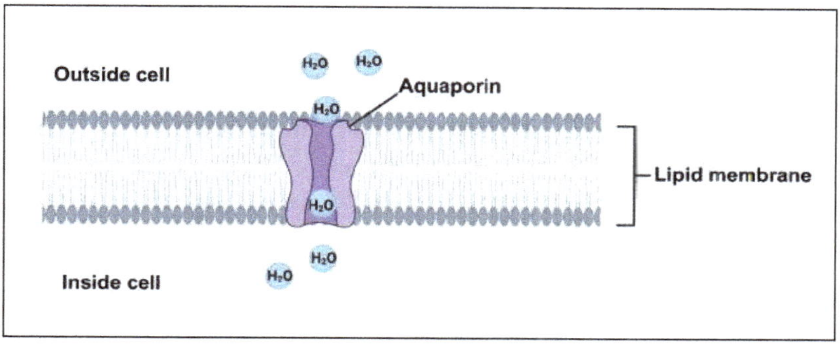

Fig. 2.15 Schematic representation of aquaporin-facilitated transmembrane movement of water molecules across membrane

2.8 Nutrient stress

Plant nutrition depends on the absorption of carbon, hydrogen and oxygen from the atmosphere and the uptake of minerals from the soil. In addition, plants require water for absorption and distribution of the elements. When these essential elements become limited in the surroundings, plants exhibit nutritional deficiencies, which is recognized as a major plant abiotic stress that dramatically affects crop yields.

Many factors contribute to nutrient losses in crop land e.g., wet soils, excess rainfall, and warming soil temperatures. These can percolate into groundwater, washed away into surface water or lost to the atmosphere, or transported below the root zone and out of reach by the plants.

Plants require 17 inorganic nutrients for optimal growth. Among these, six elements including nitrogen, phosphorus, potassium, magnesium, calcium, and sulfur are called macronutrients because they are needed in relatively large amounts by plants. Micronutrients (iron, boron, chlorine, manganese, zinc, copper, molybdenum, and nickel) are those that are needed in lesser amounts. Nitrogen, one of the major macronutrients, is typically limited in soil. Plants use nitrogen for synthesizing amino acids, proteins, nucleotides, nucleic acids, chlorophyll, and coenzymes. Thus, its deficiency can affect numerous functions leading to symptoms such as chlorosis (yellowing of leaves), and in severe cases plant death. Phosphorous, the second most important nutrient for plant growth, is usually either present as an insoluble precipitate (as calcium phosphate in alkaline soils) or sorbed in acidic soils, making it unavailable for roots to take up. Phosphorous is an important component of both cell membranes and nucleic acids and is required for all phosphorylation reactions. The remaining macronutrients, i.e., potassium, magnesium, calcium and sulfur required for a myriad of functions including but not limited to protein synthesis, activation of enzymes involved in photosynthesis, cell wall synthesis, and nucleic acid synthesis, can often be limited in soils as a consequence of low supply and leaching losses. Certain soil conditions, for instance acidic soils, can further limit the availability of micronutrients. Deficiency symptoms in plants can vary depending on the mobility of nutrients within the plant. For mobile nutrients such as nitrogen, phosphorus, potassium, and magnesium, deficiency symptoms typically appear first in older, mature leaves. This is because these nutrients can be readily translocated from older leaves to younger growing tissues when they are in short supply. Therefore, the older leaves show deficiency symptoms as they sacrifice their nutrients to support the new growth. For immobile nutrients such as calcium, sulfur, and micronutrients (e.g., iron, manganese, zinc, copper, and boron), deficiency symptoms appear first in younger tissues and new growth. This is because these nutrients cannot be easily translocated from older to younger tissues. Nutrient deficiencies in plants can cause multiple symptoms. Indeed, the nature of these symptoms can vary depending on the specific nutrient that is lacking. Fig. 2.16 illustrates a corn leaf's response to different nutrient deficiencies. Other common symptoms indicative of nutritional deficiencies include leaf necrosis, stunted growth, premature falling of leaves and delayed flowering.

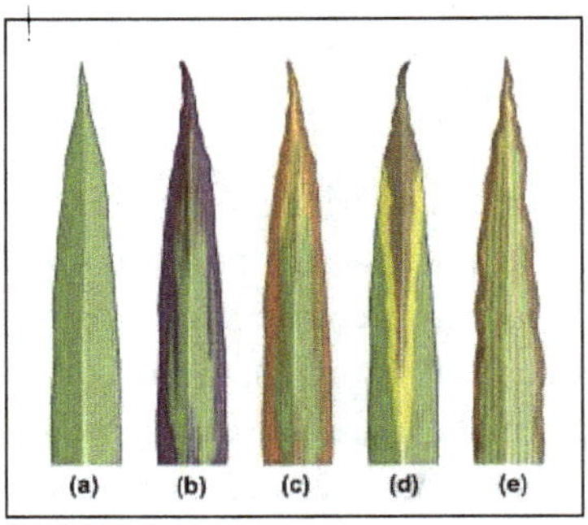

Fig. 2.16 Schematic of nutrient deficiencies in corn leaves. (a) healthy leaf (b) phosphorous deficient leaf which appears as reddish purple (c) potassium deficient leaf which appears as a firing orange or drying along the tips and edges (d) nitrogen deficient leaf which appears as yellowing (e) magnesium deficient leaf which appears as whitish lines along the veins and sometimes a purplish color on the underside of the lower leaves

Physiological and molecular mechanisms underlying crop adaptation to mineral nutrient deficiency have been well documented. For example, a morphological adaptation such as changes in root architecture promoting lateral root growth, root hair formation, clustered root formation and modification of primary root growth can improve exploration of nutrient resources in soil. In fact, plants can enhance nutrient cycling by shedding their leaves prematurely so that nutrients are allocated to areas of new growth. Changes in soluble sugars, amino acids and antioxidant enzyme levels can also suppress nutrient stress, however severe deficiency of key nutrients can hinder plant growth and reduce yield.

Nutrient management can help with some challenges. This is generally achieved through the application of chemical fertilizers in soil. Indeed, use of chemical fertilizers has increased exponentially over recent decades and could increase to 263 million tonnes by the year 2050 (Fig. 2.17).

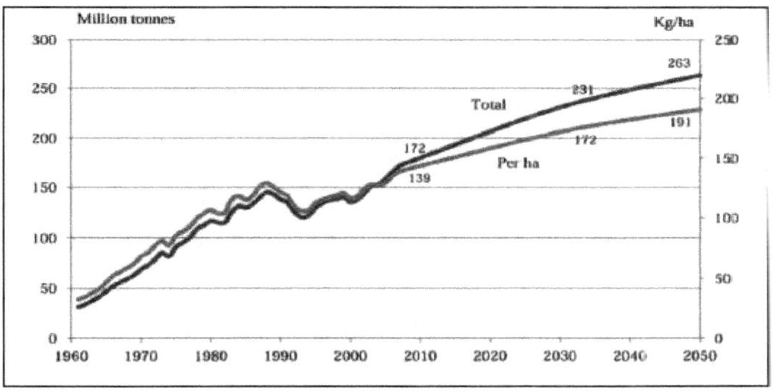

Fig. 2.17 World fertilizer consumption: past and projected

A major shortcoming of most nitrogen and phosphorous fertilizers is their high solubility in the soil. This means that much of this chemical fertilizer that is applied is lost in the environment, as plants typically absorb nutrients at a slower rate. As a result, significant nitrogen and phosphorous are lost through leaching, runoff, and/or volatilization, leading to increased environmental risks such as groundwater contamination and eutrophication (the run-off of nutrients from fields into bodies of water where these nutrients are likely to cause over-growth of algae that utilizes most of the available oxygen resulting in the widespread death of plants and animals). In severe cases, these algal blooms lead to dead zones, which are large coastal areas of an ocean that is so hypoxic that most marine life cannot survive. One of the largest dead zones is the Gulf of Mexico dead zone, situated in the USA.

Most plants form partnerships with microorganisms enabling them to deal with nutrient stresses. For instance, mycorrhizae (from the Greek words *Mycos*-fungus and *rhiza*-root) are fungi that can comprise 80% of the effective root system of a plant. As the roots penetrate the soil, the fungal hyphae can penetrate smaller pores in the soil, thereby distributing roots to a much broader area. Thus, plants acquire an efficient nutrient absorbing mechanism in which the plant provides the fungus with sugar and the fungus enables the plant to absorb more water and phosphate and other minerals than it could on its own. Those plants with a mycorrhizal symbiosis tend to be more tolerant to stressful conditions than those without. Generally, trees and shrubs are associated with ectomycorrhizae (ecto-outside, meaning those fungi that surround the roots). Endomycorrhizae or arbuscular mycorrhizae (endo-within, meaning those that live within root cells) have a broader host range and form associations with both wild plant species and agricultural crops.

Plants also associate with bacteria, for instance certain nitrogen-fixing soil bacteria (rhizobia) are able to symbiotically interact with leguminous plants (soybeans, peas, peanuts) to form nodules, specialized structures where atmospheric nitrogen gas is converted into ammonia. The bacteria supply much of the nitrogen needed by the plant and in return obtain carbohydrates from the plant. Nitrogen fixation is an energy-intensive reaction mediated by the enzyme complex nitrogenase, which is sensitive to oxygen. So, the nodule provides an oxygen free environment for the bacteria to carry out nitrogen fixation efficiently. Although this symbiosis is tightly regulated, plants often recruit other bacteria from the soil to facilitate nutrient uptake and for various other beneficial interactions. These are called plant growth-promoting bacteria (PGPB), which, unlike rhizobia that primarily focus on nitrogen fixation, enhance the availability of a broader range of essential nutrients, including phosphorus, iron, and potassium. Numerous PGPB have the ability to fix atmospheric nitrogen, solubilize phosphates and iron from soil minerals, making these nutrients more accessible to plants, and they can also enhance potassium availability through various mechanisms. Many PGPB are also involved in modulating the levels of phytohormones present in plants. Given their positive effects on nutrient management, PGPB serve as a potential source of biofertilizer and might be valuable in agriculture.

2.9 Further readings

Adeleke, Bartholomew Saanu, Saheed Adekunle Akinola, Af Adedayo, Bernard R. Glick, and Olubukola Oluranti Babalola. 2022. "Synergistic Relationship of Endophyte-Nanomaterials to Alleviate Abiotic Stress in Plants." In *Frontiers in Environmental Science*. https://api.semanticscholar.org/CorpusID:253841964.

Ahuja, Ishita, Ralph Kissen, and Atle M. Bones. 2012. "Phytoalexins in Defense against Pathogens." *Trends in Plant Science* 17 (2): 73–90. https://doi.org/10.1016/j.tplants.2011.11.002.

Alexandros, Nikos and Jella Bruinsma. 2012. World agriculture towards 2030/2050. A report by Global Perspectives Study Team. ESA Working Paper No.12-03. Agricultural development economics division. Food and Agricultural Organization of the USA.

Ali, Md Arshad, Temoor Ahmed, Ezzeldin Ibrahim, Muhammad Rizwan, Khim Phin Chong, and Jean Wan Hong Yong. 2024. "A Review on Mechanisms and Prospects of Endophytic Bacteria in Biocontrol of Plant Pathogenic Fungi and Their Plant Growth-Promoting Activities." *Heliyon* 10 (11): e31573. https://doi.org/10.1016/j.heliyon.2024.e31573.

Bailey-Serres, J., and L. A. C. J. Voesenek. 2008. "Flooding Stress: Acclimations and Genetic Diversity." *Annual Review of Plant Biology* 59:313–39. https://doi.org/10.1146/annurev.arplant.59.032607.092752.

Baxter, Aaron, Ron Mittler, and Nobuhiro Suzuki. 2014. "ROS as Key Players in Plant Stress Signaling." *Journal of Experimental Botany* 65 (5): 1229–40. https://doi.org/10.1093/jxb/ert375.

Bita, Craita, and Tom Gerats. 2013. "Plant Tolerance to High Temperature in a Changing Environment: Scientific Fundamentals and Production of Heat Stress-Tolerant Crops." *Frontiers in Plant Science* 4. https://doi.org/10.3389/fpls.2013.00273.

Bueno, Milagros, and María D. Cordovilla. 2021. "Spermidine Pretreatments Mitigate the Effects of Saline Stress by Improving Growth and Saline Excretion in Frankenia Pulverulenta." *Agronomy* 11 (8). https://doi.org/10.3390/agronomy11081515.

Chauhan, Ratna, Sanjay Sarjerao Kale, Ratan Singh, Dheeraj Rathore, Bernard R. Glick, and Indra Jeet Chaudhary. 2025. "Nature Base Sustainable Solution for Heavy Metals Toxicity and Its Impacts on Humans and Plants Health." In *Groundwater Resource Management Planning Strategies*, edited by Vangala Sunitha, Bandi Muralidhara Reddy, Yenugu Sudharshan Reddy, Mannala Prasad, Badapalli Pradeep Kumar, and Etikala Balaji. Springer Water. Springer Nature Switzerland. https://doi.org/10.1007/978-3-031-88870-0_4.

Chislock, Michael F., Enrique Doster, Rachel A. Zitomer, and Alan E. Wilson. "Eutrophication: causes, consequences, and controls in aquatic ecosystems." *Nature Education Knowledge* 4, no. 4 (2013): 10.

Conde, Artur, Paulo Silva, Alice Agasse, Carlos Conde, and Hernâni Gerós. 2011. "Mannitol Transport and Mannitol Dehydrogenase Activities Are Coordinated in Olea Europaea under Salt and Osmotic Stresses." *Plant & Cell Physiology* 52 (10): 1766–75. https://doi.org/10.1093/pcp/pcr121.

Conde, Artur, Paulo Silva, Alice Agasse, Carlos Conde, and Hernâni Gerós. 2011. "Mannitol Transport and Mannitol Dehydrogenase Activities Are Coordinated in Olea Europaea under Salt and Osmotic Stresses." *Plant & Cell Physiology* 52 (10): 1766–75. https://doi.org/10.1093/pcp/pcr121.

Cruz de Carvalho, Maria Helena. 2008. "Drought Stress and Reactive Oxygen Species: Production, Scavenging and Signaling." *Plant Signaling & Behavior* 3 (3): 156–65. https://doi.org/10.4161/psb.3.3.5536.

Cui, Jiamin, Fanxuan Nie, Yuquan Zhao, Dawei Zhang, Dinggang Zhou, Jinfeng Wu, Liang Qu, Lu Xiao, and Lili Liu. 2024. "A Review on Plant Endophytes in Response to Abiotic Stress." *Environmental Pollutants and Bioavailability* 36 (1): 2323123. https://doi.org/10.1080/26395940.2024.2323123.

Etesami, Hassan, and Bernard R. Glick. 2020. "Halotolerant Plant Growth–Promoting Bacteria: Prospects for Alleviating Salinity Stress in Plants." *Environmental and Experimental Botany* 178 (October): 104124. https://doi.org/10.1016/j.envexpbot.2020.104124.

Forni, Cinzia, Daiana Duca, and Bernard R. Glick. 2017. "Mechanisms of Plant Response to Salt and Drought Stress and Their Alteration by Rhizobacteria." *Plant and Soil* 410 (1–2): 335–56. https://doi.org/10.1007/s11104-016-3007-x.

Francis, Bonia, C.T. Aravindakumar, Philip B. Brewer, and Sibu Simon. 2023. "Plant Nutrient Stress Adaptation: A Prospect for Fertilizer Limited Agriculture." *Environmental and Experimental Botany* 213 (September):105431. https://doi.org/10.1016/j.envexpbot.2023.105431.

Gamalero, Elisa, and Bernard R. Glick. 2022. "Recent Advances in Bacterial Amelioration of Plant Drought and Salt Stress." *Biology* 11 (3): 437. https://doi.org/10.3390/biology11030437.

Gamalero, Elisa, Guido Lingua, Graziella Berta, and Bernard R. Glick. 2009. "Beneficial Role of Plant Growth Promoting Bacteria and Arbuscular Mycorrhizal Fungi on Plant Responses to Heavy Metal Stress." *Canadian Journal of Microbiology* 55 (5): 501–14. https://doi.org/10.1139/W09-010.

Gepstein, Shimon, and Bernard R. Glick. 2013. "Strategies to Ameliorate Abiotic Stress-Induced Plant Senescence." *Plant Molecular Biology* 82 (6): 623–33. https://doi.org/10.1007/s11103-013-0038-z.

Gill, Sarvajeet Singh, and Narendra Tuteja. 2010. "Reactive Oxygen Species and Antioxidant Machinery in Abiotic Stress Tolerance in Crop Plants." *Plant Physiology and Biochemistry* 48 (12): 909–30. https://doi.org/10.1016/j.plaphy.2010.08.016.

Glick, Bernard R. 2004. "Bacterial ACC Deaminase and the Alleviation of Plant Stress." In *Advances in Applied Microbiology*, vol. 56. Elsevier. https://doi.org/10.1016/S0065-2164(04)56009-4.

Glick, Bernard R. 2020. *Beneficial Plant-Bacterial Interactions*. 2nd Edn., Cham: Springer International Publishing. https://doi.org/10.1007/978-3-030-44368-9.

Hansen, J., M. Sato, R. Ruedy, K. Lo, D.W. Lea, and M. Medina-Elizade, 2006: Global temperature change. Proc. Natl. Acad. Sci., 103, 14288-14293, doi:10.1073/pnas.0606291103. n.d. "Global Temperature Change." https://doi.org/Hansen, J., M. Sato, R. Ruedy, K. Lo, D.W. Lea, and M. Medina-Elizade, 2006: Global temperature change. Proc. Natl. Acad. Sci., 103, 14288-14293, doi:10.1073/pnas.0606291103.

Hasanuzzaman, Mirza, M. H. M. Borhannuddin Bhuyan, Taufika Islam Anee, Khursheda Parvin, Kamrun Nahar, Jubayer Al Mahmud, and Masayuki

Fujita. 2019a. "Regulation of Ascorbate-Glutathione Pathway in Mitigating Oxidative Damage in Plants under Abiotic Stress." *Antioxidants (Basel, Switzerland)* 8 (9). https://doi.org/10.3390/antiox8090384.

Kerbler, Sandra M., and Philip A. Wigge. 2023. "Temperature Sensing in Plants." *Annual Review of Plant Biology* 74 (May):341–66. https://doi.org/10.1146/annurev-arplant-102820-102235.

Khan, Zareen, Grant Guelich, Ha Phan, Regina Redman, and Sharon Doty. 2012. "Bacterial and Yeast Endophytes from Poplar and Willow Promote Growth in Crop Plants and Grasses." *ISRN Agronomy* 2012 (October):1–11. https://doi.org/10.5402/2012/890280.

Khan, Zareen, Shyam Kandel, Daniela Ramos, Gregory Ettl, Soo-Hyung Kim, and Sharon Doty. 2015. "Increased Biomass of Nursery-Grown Douglas-Fir Seedlings upon Inoculation with Diazotrophic Endophytic Consortia." *Forests* 6 (12): 3582–93. https://doi.org/10.3390/f6103582.

Khan, Zareen, and Sharon L. Doty. 2009. "Characterization of Bacterial Endophytes of Sweet Potato Plants." *Plant and Soil* 322 (1–2): 197–207. https://doi.org/10.1007/s11104-009-9908-1.

Krishnan-Schmieden, Maithili, Patrick E. Konold, John T. M. Kennis, and Anjali Pandit. 2021. "The Molecular pH-Response Mechanism of the Plant Light-Stress Sensor PsbS." *Nature Communications* 12 (1): 2291. https://doi.org/10.1038/s41467-021-22530-4.

Levitt, J. 1980. "Responses of Plants to Environmental Stress." In *Chilling, Freezing, and High Temperature Stresses. Academic Press, Cambridge.* Vol. Volume 1. Academic Press, Cambridge. https://doi.org/10.1016/B978-0-12-445501-6.50016-6.

Miller, Gad, Nobuhiro Suzuki, Sultan Ciftci-Yilmaz, and Ron Mittler. 2010. "Reactive Oxygen Species Homeostasis and Signaling during Drought and Salinity Stresses." *Plant, Cell & Environment* 33 (4): 453–67. https://doi.org/10.1111/j.1365-3040.2009.02041.x.

Mittler, Ron, Sandy Vanderauwera, Nobuhiro Suzuki, Gad Miller, Vanesa B. Tognetti, Klaas Vandepoele, Marty Gollery, Vladimir Shulaev, and Frank Van

Breusegem. 2011. "ROS Signaling: The New Wave?" *Trends in Plant Science* 16 (6): 300–309. https://doi.org/10.1016/j.tplants.2011.03.007.

Mohapatra, S. S., R. J. Poole, and R. S. Dhindsa. 1988. "Abscisic Acid-Regulated Gene Expression in Relation to Freezing Tolerance in Alfalfa." *Plant Physiology* 87 (2): 468–73. https://doi.org/10.1104/pp.87.2.468.

Muhammad, Murad, Abdul Waheed, Abdul Wahab, Muhammad Majeed, Muhammad Nazim, Yong-Hong Liu, Li Li, and Wen-Jun Li. 2024. "Soil Salinity and Drought Tolerance: An Evaluation of Plant Growth, Productivity, Microbial Diversity, and Amelioration Strategies." *Plant Stress* 11 (March):100319. https://doi.org/10.1016/j.stress.2023.100319.

Munns, Rana, and Mark Tester. 2008. "Mechanisms of Salinity Tolerance." *Annual Review of Plant Biology*. Annual Reviews. https://doi.org/10.1146/annurev.arplant.59.032607.092911.

Oyedoh, Oghoye Priscilla, Anjana J. Atapattu, Jing Ge, Gustavo Santoyo, Bernard R. Glick, and Olubukola Oluranti Babalola. 2024. "Biotic Stress Alleviation: A Sneak Peek into the Beneficial Taxa in Rhizosphere." *Biocatalysis and Agricultural Biotechnology* 62 (December): 103427. https://doi.org/10.1016/j.bcab.2024.103427.

Patel, Margi, Shaikhul Islam, Bernard R. Glick, Shobhit Raj Vimal, Sachin Ashok Bhor, Matteo Bernardi, Fatema Tuj Johora, Ashish Patel, and Sergio de los Santos Villalobos. 2024. "Elaborating the Multifarious Role of PGPB for Sustainable Food Security under Changing Climate Conditions." *Microbiological Research* 289 (December):127895. https://doi.org/10.1016/j.micres.2024.127895.

Pearce, Roger S. 2001. "Plant Freezing and Damage." *Annals of Botany* 87 (4): 417–24. https://doi.org/10.1006/anbo.2000.1352.

Ristaino, Jean B., Pamela K. Anderson, Daniel P. Bebber, Kate A. Brauman, Nik J. Cunniffe, Nina V. Fedoroff, Cambria Finegold, et al. 2021. "The Persistent Threat of Emerging Plant Disease Pandemics to Global Food Security." *Proceedings of the National Academy of Sciences* 118 (23): e2022239118. https://doi.org/10.1073/pnas.2022239118.

Santos, Tiago Benedito dos, Alessandra Ferreira Ribas, Silvia Graciele Hülse de Souza, Ilara Gabriela Frasson Budzinski, and Douglas Silva Domingues. 2022. "Physiological Responses to Drought, Salinity, and Heat Stress in Plants: A Review." *Stresses* 2 (1): 113–35. https://doi.org/10.3390/stresses2010009.

Gustavo Santoyo, Elisa Gamalero, and Bernard R. Glick. 2021. "Mycorrhizal-Bacterial Amelioration of Plant Abiotic and Biotic Stress." *Frontiers in Sustainable Food Systems* 5. https://doi.org/10.3389/fsufs.2021.672881.

Shafiq, Iram, Sajad Hussain, Muhammad Ali Raza, Nasir Iqbal, Muhammad Ahsan Asghar, Ali Raza, Yuan-fang Fan, et al. 2021. "Crop Photosynthetic Response to Light Quality and Light Intensity." *Journal of Integrative Agriculture* 20 (1): 4–23. https://doi.org/10.1016/S2095-3119(20)63227-0.

Shi, Yiting, Yanglin Ding, and Shuhua Yang. 2018. "Molecular Regulation of CBF Signaling in Cold Acclimation." *Trends in Plant Science* 23 (7): 623–37. https://doi.org/10.1016/j.tplants.2018.04.002.

Silva, Paulo, Arnoldo R. Façanha, Rui M. Tavares, and Hernâni Gerós. 2010. "Role of Tonoplast Proton Pumps and Na+/H+ Antiport System in Salt Tolerance of Populus Euphratica Oliv." *Journal of Plant Growth Regulation* 29 (1): 23–34. https://doi.org/10.1007/s00344-009-9110-y.

Skuza, Lidia, Izabela Szućko-Kociuba, Ewa Filip, and Izabela Bożek. 2022. "Natural Molecular Mechanisms of Plant Hyperaccumulation and Hypertolerance towards Heavy Metals." *International Journal of Molecular Sciences* 23 (16). https://doi.org/10.3390/ijms23169335.

Thomas, Dhanya T. T., Dinakar Challabathula, and Jos T. Puthur. 2019. "UV-B Priming of Oryza Sativa Var. Kanchana Seedlings Augments Its Antioxidative Potential and Gene Expression of Stress-Response Proteins under Various Abiotic Stresses." *3 Biotech* 9 (10): 375. https://doi.org/10.1007/s13205-019-1903-5.

Thomashow, Michael F. 1998. "Role of Cold-Responsive Genes in Plant Freezing Tolerance1." *Plant Physiology* 118 (1): 1–8. https://doi.org/10.1104/pp.118.1.1.

Valverde, Roberto, Tony H. H. Chen, and Paul H. Li. 1997. "Frost Hardiness and Cold Acclimation in Solanum Species." In *Plant Cold Hardiness: Molecular Biology, Biochemistry, and Physiology*, edited by Paul H. Li and Tony H. H. Chen, 57–66. Boston, MA: Springer US. https://doi.org/10.1007/978-1-4899-0277-1_6.

Westwood, MN. 1978. *Temperate-Zone Pomology. Temperate-Zone Pomology.* W.H. Freeman and Company.

Williamson, V M, and R S Hussey. 1996. "Nematode Pathogenesis and Resistance in Plants." *The Plant Cell* 8 (10): 1735–45. https://doi.org/10.1105/tpc.8.10.1735.

Zhang, Yan, Jing Xu, Ruofan Li, Yanrui Ge, Yufei Li, and Ruili Li. 2023. "Plants' Response to Abiotic Stress: Mechanisms and Strategies." *International Journal of Molecular Sciences* 24 (13). https://doi.org/10.3390/ijms241310915.

Zhu, Jian-Kang. 2016. "Abiotic Stress Signaling and Responses in Plants." *Cell* 167 (2): 313–24. https://doi.org/10.1016/j.cell.2016.08.029.

Chapter 3

Plant Molecular Biology

Most plants have a very large genome, typically several times the size of most animal genomes (Table 3.1). In addition, many plants are polyploids; that is, they contain more than two complete sets of chromosomes (i.e., one from each parent). Some highly cultivated plants such as wheat have some strains that are diploid (containing two sets of chromosomes), some strains that are tetraploid (containing four sets of chromosomes), and some that are hexaploid (containing six sets of chromosomes). Plants with different ploidy levels usually cannot interbreed.

Plant plastids and mitochondria are believed to have originated from endosymbiotic bacteria. However, over time these organelles have lost most of their original DNA (in many instances this lost DNA can be found as part of the plant nuclear DNA). Thus, plant plastids generally contain ~120-160 kb of double stranded DNA while plant mitochondria typically contain ~200-2600 kb of double stranded DNA.

3.1 The plant genome

A number of genomes from various organisms have been sequenced, as indicated in Table 3.1. Compared with other organisms, plant genomes tend to be larger and more complex, and plants often contain high levels of polyploidy (discussed in the next section) and repeated DNA sequences. Larger genomes can impose constraints on many functional traits in plants, including a slower growth rate and longer developmental cycle. Thus, *Arabidopsis thaliana* exhibits both rapid growth and development, making it a model plant for studying plant biology, biochemistry and genetics. Moreover, it is much easier to isolate and characterize mutants of plants with smaller genomes. Further comparisons of genomes reveal that, despite significant size differences, it is unreasonable to conclude that the large variation in genome size alone underlies the differences in biological complexity between organisms. For example, comparing a pea plant and *E. coli*; although the pea has a much larger genome than *E. coli*, this difference does not directly translate to greater biological complexity. Instead, factors such as the regulation of gene expression and the presence of coding sequences play crucial roles in determining an organism's complexity. Thus, it is quite common for only a relatively small portion of a plant's genome to be composed of protein coding genes. For example, with pea plants, only about 4% of the total DNA contains codes for proteins, the remaining DNA includes non-coding regions and repeated sequences that are repeated multiple times in the genome. Also, unlike bacteria where most genes are present in only a single copy, numerous plant genes are present in multiple copies (and in some cases the same gene may be regulated under the control of different promoters).

Table 3.1 Range of genome sizes in higher plants compared to a virus, a bacterium, a strain of yeast, an animal cell and a human cell. The range in genome size that is indicated for any particular plant represents differences in various subspecies of the plant

Organism	Scientific name	Genome size (millions of base pairs, Mb)
Virus	*Cauliflower mosaic virus (CaMV)*	0.008
Bacteria	*Escherichia coli*	4.6
Brewer's yeast	*Saccharomyces cerevisiae*	14
Mouse	*Mus musculus*	2,700
Human	*Homo sapiens*	3,117
Pea	*Lathyrus oleraceus*	4,300
Corn	*Zea mays*	2,292-2,716
Papaya	*Carica papaya*	372
Rice	*Oryza sativa*	419-463
Tobacco	*Nicotiana tabacum*	4,221-4,646
Peanut	*Arachis hypogaea*	2,813
Soybean	*Glycine max*	1,115
Tomato	*Lycopersicum esculentum*	907-1,000
Bread wheat	*Triticum aestivum*	15,966
Thale cress	*Arabidopsis thaliana*	0.145

The location of the genes within chromosomes, their proximity to regulatory elements and the chromatin conformation all affect the extent to which the genes are available for transcription (into mRNA). Plant development and differentiation involve the selective expression of specific genes at different stages of plant growth and development. In fact, there is a whole range of different DNA sequences and factors that function to determine when and where any particular gene will be expressed. These regulatory sequences may be broadly classified into simple promoters, enhancers and regulatory regions. Promoters are often found at a fixed characteristic position at the beginning of genes. Enhancers are DNA sequences that can augment gene expression. They can be located several hundred kilobases away from the target gene and are position orientation dependent. Regulatory sequences are DNA sequences which induce and control gene expression. For example, in some cases, methylation at specific sites may be important in regulating gene expression.

3.2 Plant polyploidy

Polyploid plants contain two or more sets of chromosomes (Table 3.2) and can occur in the wild or may be cultivated by artificial means (e.g., treatment with chemicals). In nature, polyploidy seems to provide a major mechanism for adaptation and genetic variation, and this feature has led to the development of crops with polyploid genomes that are useful for application in breeding programs. Perhaps, one of the most remarkable examples has been the exploitation of polyploid nature of wheat to improve its agricultural traits, such as yield, disease resistance, and adaptability to different environments. Wheat is a complex polyploid, with common bread wheat (*Triticum aestivum*) being a hexaploid species (2n = 6x= 42 chromosomes), meaning that wheat has six sets of chromosomes that are derived from three different ancestral species.

Table 3.2 Examples of polyploid crops

Common name	Scientific name	Valuable market trait	Ploidy	Chromosome number
Potato	*Solanum tuberosum*	Edible tuber	4x	48
Strawberry	*Fragaria × ananassa*	Edible fruit	8x	56
Banana	*Musa acuminata*	Edible fruit	3x	33
Sweet potato	*Ipomea batas*	Edible tuber	6x	90
Cotton	*Gossypium hirsutum*	Usable fiber	4x	52
Coffee	*Coffea arabica*	Beverage	4x	44
Durum wheat	*Triticum turgidum L.*	Grain	4x	28
Watermelon	*Citrullus vulgaris*	Edible fruit	3x	33
Oat	*Avena sativa* L.	Grain	6x	42
Sugar beet	*Beta vulgaris L.*	Industrial	6x	27

Polyploidy increases genome size, and a significant consequence of becoming a polyploid is an increase in plant cell size compared to their diploid relative, leading, in many cases, to enlarged plant organs (Fig. 3.1).

Fig. 3.1 Polyploid plants (shown on the right) of Arabidopsis (top) and sweet orange (bottom) compared to their diploid counterparts (shown on the left)

Typically, polyploids are more robust plants that are taller with larger flowers, seeds and fruit. In general, these plants have a slower growth rate resulting in a prolonged or later onset of flowering, which might be a useful trait in ornamental breeding. Apart from morphological changes, polyploidy can also alter the quality and quantity of secondary metabolites in plants. Several aromatic polyploids, such as chamomile (*Matricaria chamomilla* L.; 2n = 18) are cultivated worldwide due to increased flavonoid content and elevated levels of essential oils.

Induced polyploids of some species appear to be more tolerant of abiotic and biotic stresses, which have made them increasingly important in plant breeding. When compared to their diploid counterparts, polyploid plants show greater resistance to cold, drought, salinity, pests and pathogens. Notably, the response varies by species, for instance some polypoid wheat varieties show a higher tolerance to nutritional stress when compared to *Solanum* species (including tomato, eggplant and potato). Polyploidy is also utilized in forage breeding, offering several advantages that enhance the both the quality and productivity of forage crops (grown to feed livestock). For example, addition of extra chromosomes (via polyploidy) in rye grass has resulted in cultivars that are highly productive and palatable, consequently enhancing animal productivity (since rye grass is primarily used as cattle feed). Moreover, polyploids can serve as an effective bridge allowing for successful hybridization between species that cannot cross directly due to differences in ploidy levels. This method is very common in apples. Thus, for instance, the tetraploid variety produces low quality apples with low resistance to cold. However, the tetraploid variety was used to develop the triploid variety which tended to have improved fruit quality, larger fruit size, and/or greater cold resistance.

There are some detrimental effects of polyploidy as well. Polyploidy is typically accompanied by a reduction in plant fertility, which can be a major limitation for seed dependent crops. Nonetheless, in cases that include seedless fruits (e.g., seedless watermelon and grapes) or sugar beets (where only roots are relevant) it is a quite beneficial trait. Moreover, for ornamental breeding, this disadvantage is offset by larger and more beautiful flowers, such as the polyploid varieties of tulips which are characterized by their large blooms.

3.3 Introns and exons

RNA processing involves many steps. Most plant genes contain regions of DNA called introns or intervening regions, that do not code for a protein. In some cases, the total length of introns considerably exceeds the length of the coding regions (called exons) within a gene. For example, in Arabidopsis and rice, the total intron size can make up 30-40% of the genome. Similarly, the number of introns per gene is highly variable. All introns are transcribed from DNA into RNA; however, the introns must be removed from the pre-mRNA (to form mRNA) before protein synthesis can occur. The process of removing introns is called splicing, and this takes place by groups of splicing enzymes (working in concert) called spliceosomes. These enzymes remove the introns and precisely ligate the coding sequences, or exons, together to generate mature mRNA. Spliceosomes are made up of large RNA-protein complexes called small nuclear ribonucleoproteins (snRNPs or 'snurps') which recognize precise sequences at the intron-exon boundaries. The mature mRNA undergoes further processing which includes capping of the mRNA at the 5' end and insertion of a poly A tail at the 3' end. These changes are necessary to maintain mRNA stability and to facilitate mRNA translation into protein (Fig. 3.2).

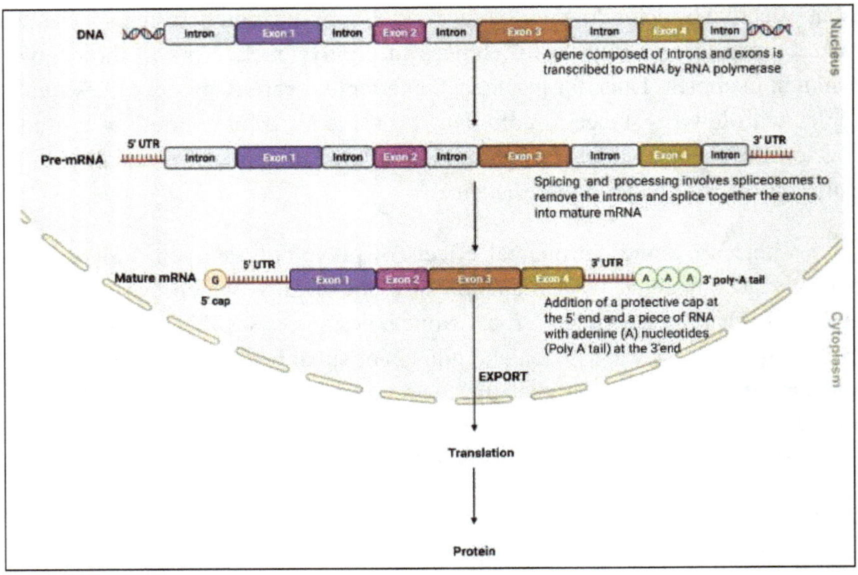

Fig. 3.2 A schematic representation of the steps of protein production. The DNA is transcribed into pre-mRNA which is then spliced to remove the introns. The exons are joined, to form a functional mRNA. The mature mRNA is capped at the 5' end and a poly A tail is added to the 3' end before the mature mRNA is released into the cytoplasm where it is translated into protein

RNA splicing can occur in a variety of ways thus affecting the gene product in the cell. The following two scenarios are possible:

(i) Alternative splicing, where spliceosomes can splice one type of pre-mRNA different ways, leading to multiple mature mRNA molecules. These variations in mRNA leads to differences in proteins ultimately producing several different gene products from one original pre-mRNA transcript (Fig. 3.3).

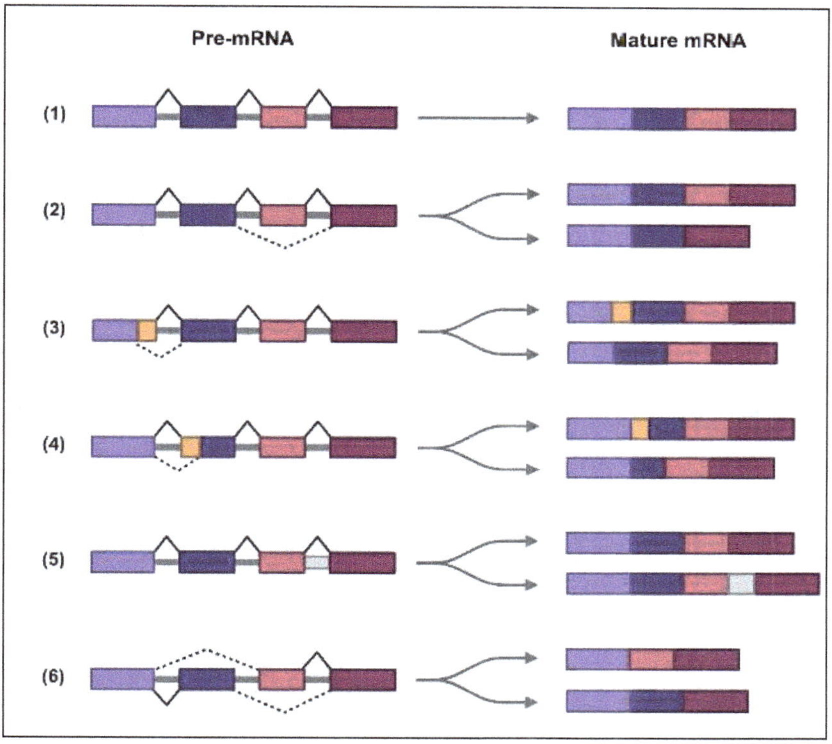

Fig. 3.3 Illustration of different splicing types by which mRNA transcripts are generated (1) constitutive splicing (2) exon skipping/inclusion (3) alternative 5' splice sites (4) alternative 3' splice sites (5) intron retention (6) mutually exclusive exons. Alternatively spliced transcripts can yield proteins with different structures or metabolic functions

RNA interference or RNAi is a mechanism to selectively degrade specific mRNAs. RNAs called small interfering RNAs (siRNA) or microRNAs (miRNA) may be formed to target mRNAs by complementarity preventing them from being translated into proteins (Fig. 3.4). The usefulness of this method is that it can block the expression of the gene that encoded the mRNA with the result that no protein is produced from it. RNAi might be used by a plant to protect itself from various plant viruses.

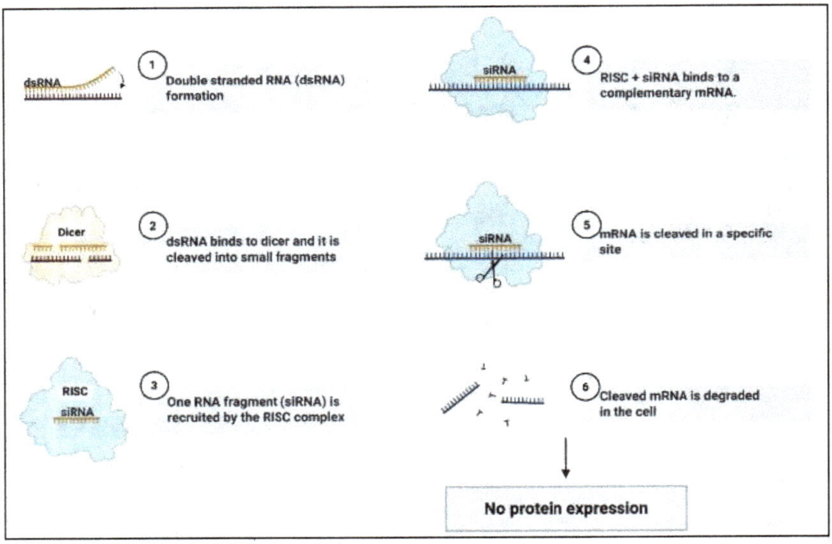

Fig. 3.4 Schematic representation of the mechanism of RNAi

3.4 Organellar DNA

According to the endosymbiotic theory, chloroplasts and mitochondria were originally derived from symbiotic bacteria. This theory, originally proposed in the mid 19th century was based upon the idea of a detailed comparison of the physical appearance of these organelles to that of bacteria. However, this idea lay dormant until the 1960's when Dr. Lynn Margulis moved the theory forward. She proposed that endosymbiosis was very important in the origin of eukaryotic cells, and that it led to the development of cell organelles, including chloroplasts and mitochondria. Although her work was initially rejected, it was accepted several years later when prokaryotic-like ribosomes and DNA were detected within both chloroplasts and mitochondria. In addition, when chloroplast and mitochondrial genomes were sequenced, and when the DNA sequences of these organelles were compared to those of bacteria, striking similarities were found. These data have led modern scientists to agree with the endosymbiotic theory of the origin of chloroplasts and mitochondria. However, over time chloroplast and mitochondrial genes and systems which were no longer necessary may have been lost or transferred to the nuclear DNA. In higher plants, for example, two-thirds of the

60 or so chloroplast ribosomal proteins are encoded in the genomic DNA within the cell's nucleus.

The chloroplast DNA in exists as multiple copies, ranging in size from 120-160 kilobase (kb) pairs, whereas mitochondrial genomes are much larger and vary in size from 200-2600 kb. There is also some evidence in the evolutionary history of some plants of gene transfer having occurred from the chloroplast genome into the mitochondrial genome. The chloroplast genome organization is very similar in all plants, however, the mitochondrial genomes in higher plants, even in closely related plant species, often have a different gene organization. Surprisingly, introns are sometimes present in chloroplast and mitochondrial genes as introns are generally quite rare in bacterial genes. Moreover, the size of the organellar genome can vary in size due to the presence or absence of introns. In some cases, the introns may be absent or in extreme cases introns may comprise the majority of the mitochondrial genome.

Chloroplast genes are involved in four main types of processes: transcription, translation, photosynthesis, and the biosynthesis of small molecules such as amino acids, fatty acids, and pigments whereas mitochondria synthesize a relatively small number of proteins including those involved in electron transport and ATP production. Organelle genes are typically inherited maternally, however in a few plants, chloroplasts can be inherited from both parents.

3.5 Transposable elements

Transposable elements (TEs) are relatively small segments of DNA that can move (transpose) from one region of a genome to another. These elements are quite common, forming a large portion of plant genomes. TEs were first described in corn (maize) by the Nobel Laureate Dr. Barbara McClintock who named them 'controlling elements' or 'jumping genes' because of their ability to move around the genome and modulate gene expression. The importance of TEs was not widely appreciated in the 1940s when they were first observed by Dr. McClintock. However, with the advent of modern molecular biology in the 1980s, her work was finally understood to be fundamentally transformative. Some types of TEs may insert themselves within genes, inactivating those genes or creating new regulatory elements. Alternatively, in some cases TEs have been shown to be excised from an otherwise inactive gene which then becomes active. Since scientists began to appreciate the role and importance of TEs, they have become useful in the study of gene regulation in plants (as well as in bacteria and animals). Some TEs are, however, quite stable and remain in the same location in the genome for long periods of time. DNA transposons are a type of TE that can move throughout the genome by a mechanism called "cut and paste", in which a piece of DNA is moved from its original location and inserted into a new location. These transposons generally contain the information for their own transposition i.e., a gene that codes for the enzyme transposase, which catalyzes the cutting and resealing of DNA and recognition sites (Fig. 3.5).

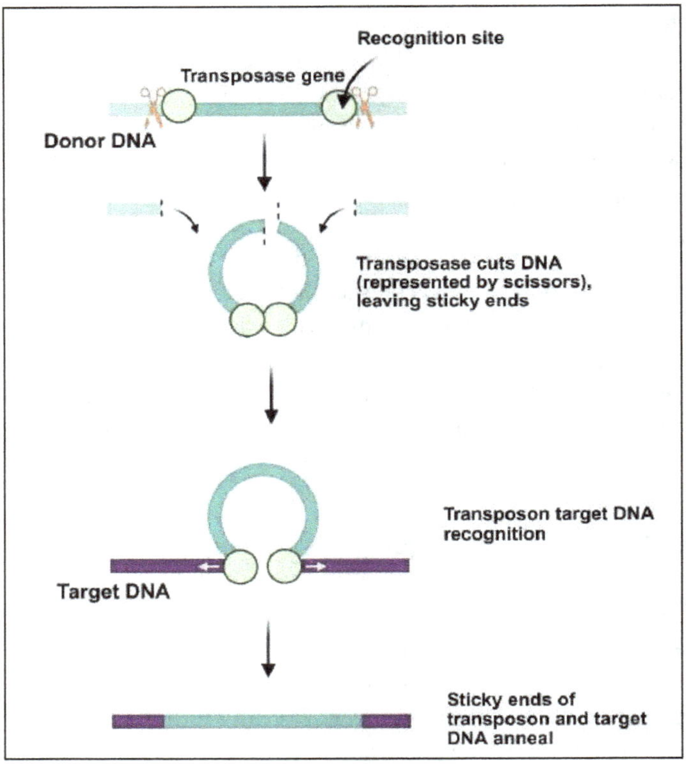

Fig. 3.5 Transposons move by directly cutting themselves out of one spot in the genome and pasting themselves into another

In contrast, retrotransposons, can change positions through a "copy and paste" mechanism and rely on a gene that codes for the enzyme reverse transcriptase for their transposition. However, there are cases where a transposon cannot be transposed by itself but requires another element to move to a new genetic location. For example, one type of maize insertion element known as *Dissociation* (DS) cannot be transposed by itself, but requires another element known as an *Activator* (AC) to move.

The capacity to move around the genome either on their own or under the influence of other genes has given TEs the name 'jumping genes'. Transposon movement can be destructive to the host genome. For example, TEs can lead to gene disruption (gene deletion or duplication), changes in gene expression and overall gene functionality of genes located near the site where TE was inserted or removed. Thus, TEs can create genome instability and can negatively affect plant growth and development. It is possible to suppress the activity of TEs by epigenetic modifications (discussed later) to maintain genome integrity.

Indeed, transposons are a common mechanism for moving genes from one chromosome to another. Moreover, because they may be carried between cells on plasmids (self-replicating circular DNA molecules carrying genes that are usually not important for the cell's survival), they can also spread from one organism, or even species, to another. For example, foreign genes may be integrated into plant cells via DNA constructs with transposon sequences, which is a well-established method in plant genetic engineering. Thus, in addition to the transposon tagging of genes, the essential features of the sequence can be used for many genetic studies.

3.6 Transcription and translation mechanisms

Transcription and translation are mechanisms by which cells express the genetic information contained in their genes to produce a protein. Cells can synthesize vast amounts of some proteins and small amounts of others due to differences in the efficiencies of transcription and translation.

We know that DNA is maintained in the cell in the form of chromatin. The loosely adhered section of chromatin is where transcription happens, where portions of a DNA sequence i.e., a gene is copied into an RNA nucleotide sequence. The mechanism of transcription is represented in Fig. 3.6.

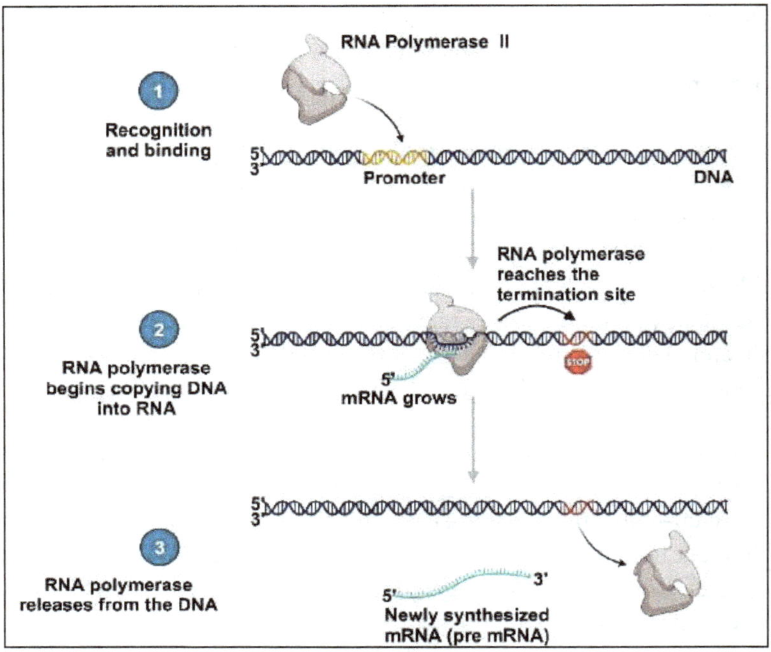

Fig. 3.6 Stages of transcription

During transcription, the DNA is unwound, and one strand is used as a template for the synthesis of a pre mRNA molecule. Many identical RNA copies can be made from the same gene, and each mRNA molecule can eventually direct the synthesis of many identical proteins. Transcription is initiated by an RNA polymerase enzyme which moves along the DNA, unwinding the double stranded molecule and begins forming an RNA chain.

In plants, there are 3 types of RNA polymerase enzymes, i.e., RNA polymerase I, II, and III. All are multi subunit complexes and similar in structure, however each transcribes a different set of genes. RNA polymerase I and III transcribe genes encoding transfer RNA, ribosomal RNA, and other small RNAs. RNA polymerase II transcribes all protein coding genes and requires transcription factors for transcription to begin. Promoter sequences in plant genes contain a common sequence, TATAAA or the 'TATA box', which is usually 15-60 nucleotides long. These sequences are located upstream of the coding region and are the site to which the RNA polymerase binds prior to transcription of the gene. Promoters do not code for proteins but play an important role in regulating gene expression by controlling the initiation, termination and abundance of transcription. Downstream of the gene, there is a termination signal i.e., a terminator for transcription to end (Fig. 3.7). Once the newly made RNA chain is released, the polymerase searches for a new promoter to restart the transcription process.

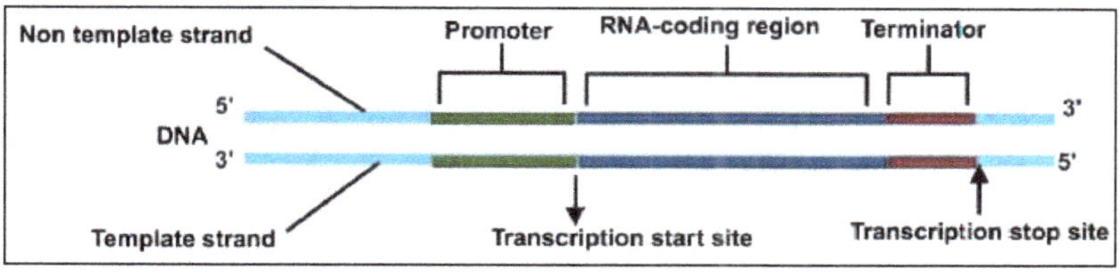

Fig. 3.7 A DNA segment showing promoter and terminator sites

The synthesized pre-mRNA (transcript) undergoes some important processing steps necessary to prevent mRNA degradation (see section 3.3), and translation can then proceed. Each transcript typically carries information for the synthesis of one protein. Thus, each single gene codes for a single mRNA which in turn codes for a single protein.

The sequence of codons (groups of three nucleotides) on the mRNA determines the sequence of amino acids in the protein being synthesized. The process of converting the codons on the mRNA to proteins is called translation. The site of translation is the ribosome, the vast majority of which in plants are the 80S type. The tRNAs which bring the amino acids to the site of synthesis on the ribosome are synthesized by RNA polymerase III and these tRNAs each carrying a specific amino acid recognize and bind both to a specific codon on the mRNA. Fig. 3.8 shows the details of translation.

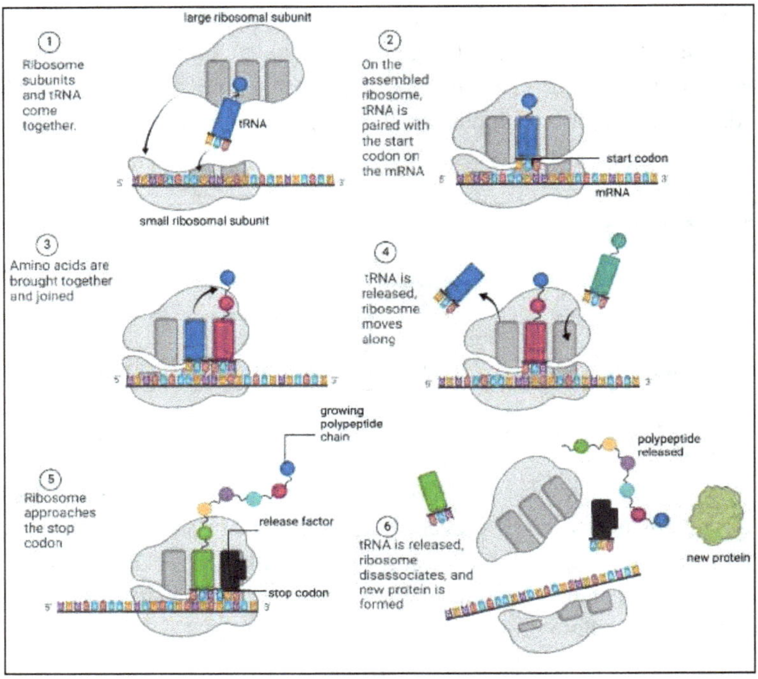

Fig. 3.8 A schematic representation of the steps of translation

Following the onset of translation, the ribosome moves along the mRNA, exposing sites (codon by codon) until one of the stop codons is reached. Next, a release factor (enzymes responsible for breaking the bond between the polypeptide chain and the tRNA) binds to the ribosome and terminates translation, in the process releasing the polypeptide chain to form a new protein. The newly synthesized proteins are often made in an inactive precursor form, i.e., a preprotein, which undergoes a series of post-translational modifications to convert it into a mature active protein. It is possible to target proteins to specific locations within a cell to improve protein stability and functionality. For example, signal or transit peptides enable the targeting of proteins to improve their synthesis, stability and even nutritional value (discussed in Chapter 5). In fact, there are many opportunities during the pathway from gene to protein for controlling gene expression in plants (Fig. 3.9).

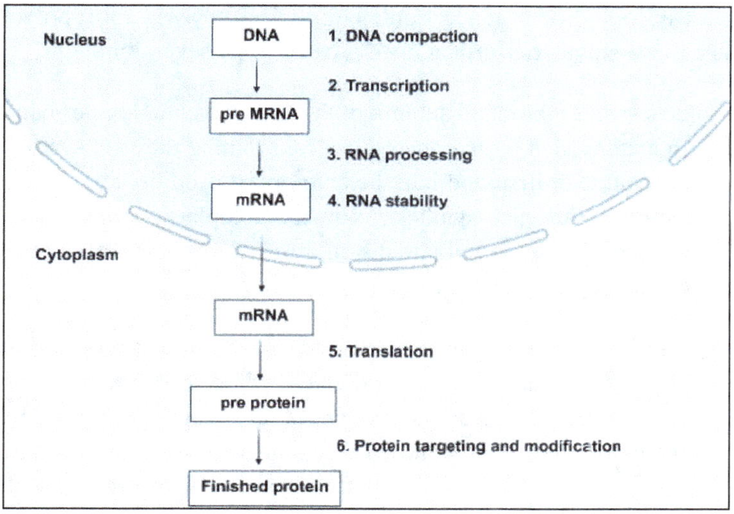

Fig. 3.9 Different points of intervention throughout the gene regulation process

3.7 Protein synthesis in the chloroplast

Chloroplasts contain the necessary genetic machinery for transcription and protein synthesis. However, most of the chloroplast proteins are encoded in the nuclear genome. For instance, the small subunit of ribulose bisphosphate carboxylase (Rubisco) is encoded by nuclear genes whereas the large subunit of Rubisco is encoded by the chloroplast DNA. Rubisco is not only a key protein in the conversion of carbon dioxide into sugar, but also one of the most abundant proteins on the planet. Protein synthesis of this enzyme is a coordinated effort of the chloroplast and the nucleus.

Two types of RNA polymerases are responsible for transcription, one is coded by the chloroplast genome, which has a strong homology with the RNA polymerases that are found in bacteria, and the other enzyme is encoded within the plant's nuclear DNA. Chloroplasts possess a bacterial type 70S ribosome (with a 50S large subunit and a 30S small subunit), as well as a complete set of bacterial type ribosomal RNAs (i.e., 23S, 16S, and 5S rRNA) with functions similar to those found in bacteria. Moreover, the mechanism of protein synthesis within chloroplasts is more similar to that found in bacteria than eukaryotes. In addition, there are some unique aspects of transcription and processing of chloroplast mRNA. For instance, some chloroplast mRNAs form polycistronic units, meaning a single transcribed mRNA can encode several proteins.

Proteins such as the small subunit of the Rubisco complex, ferredoxin, and plastocyanin are encoded by the nuclear genes, meaning they are synthesized in the cytoplasm and then imported into the chloroplast. Additionally, several other chloroplast proteins are also synthesized in the cytoplasm in the form of pre-proteins and need to be imported into the chloroplast for their proper functioning and the overall process of photosynthesis.

Typically, most preproteins synthesized in the cytoplasm but destined for the chloroplast contain additional amino acids linked to the N-terminus of the mature proteins, which act as signal sequences for recognition by the receptors located in the chloroplast membranes. These signals, also known as N-terminal chloroplast targeting peptides (cTPs) or transit peptides, are recognized by the protein complexes (receptor proteins and channel proteins) that form channels facilitating the import of preproteins to their functional location inside the chloroplast. Upon import, the transit peptide is cleaved from the preprotein and degraded, resulting in a mature functional protein within the chloroplast (Fig. 3.10). This reaction is performed by specific peptides that the chloroplast proteins recognize. Since the N-terminal sequence is cleaved upon import, this mechanism is useful to direct foreign proteins to the chloroplast (Chapter 5).

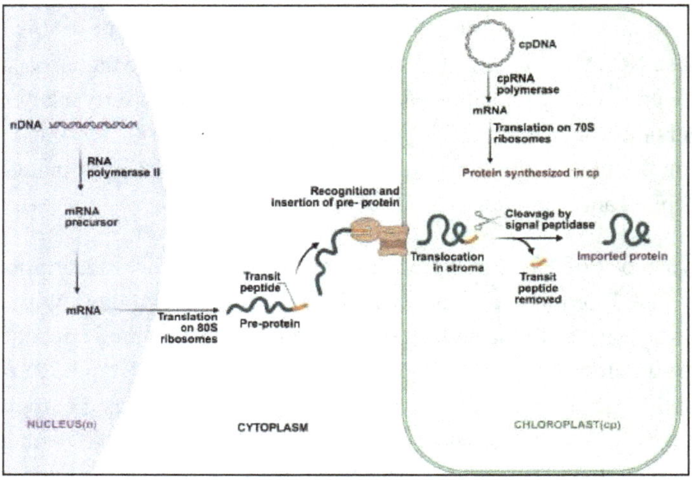

Fig. 3.10 Mode of translocation of proteins into the chloroplast by the help of signal peptides. Abbreviations: cp-chloroplast; n-nucleus

3.8 DNA methylation

Gene expression is a highly regulated process, and not all genes are active at all times. Changes to the DNA can affect gene expression, which in turn influences protein production. Cells can silence genes by attaching methyl groups to portions of the DNA, causing the chromatin to become more compact and rigid, while making a particular gene less accessible for transcription (Fig. 3.11). This is a form of epigenetic modification called DNA methylation, a mechanism which alters gene expression by modifying the DNA without changing the DNA sequence itself. It is, however, a heritable change. The methylated (off) genes are passed to offspring cells although these genes may be turned on in a later generation; this is called epigenetic inheritance.

Methylation in human and animal genomes typically occurs on cytosine bases of the DNA at a particular type of location. However, plants methylate cytosines at more positions: CG (a cytosine followed by guanine), CHG (where H corresponds to A, T, or C), and CHH.

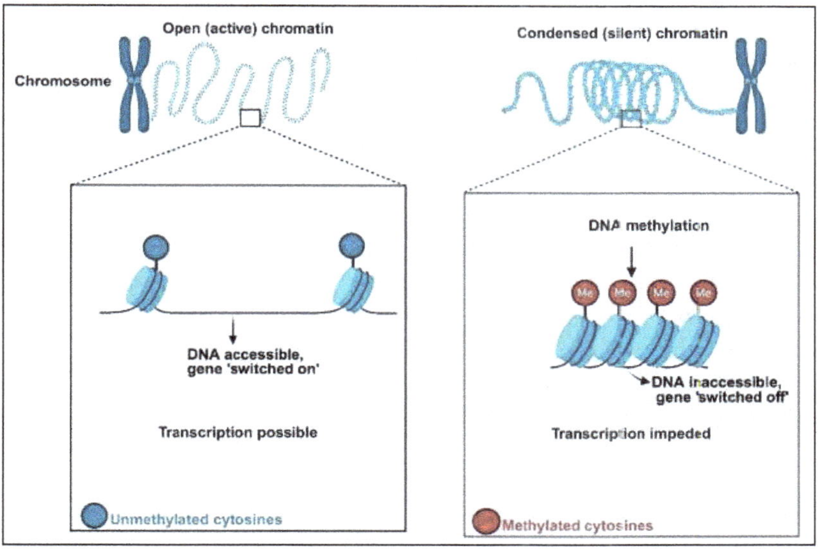

Fig. 3.11 Cells can silence genes by adding methyl groups to portions of the DNA (shown as red circles), causing the chromatin to become compact and therefore gene expression is blocked

DNA methylation affects various biological processes in plant cells in different ways. Such changes can be triggered by external stimuli affecting many molecular mechanisms within plant cells. Often, the promoter region is targeted for methylation which represses transcription and gene expression. However, methylation is not just limited to the promoter regions of the genes, but also to their coding regions, impacting transcript elongation. Changes of DNA methylation can also occur in transposable elements. By adding a methyl group, transposons can be 'silenced' taking a role in maintaining plant genome stability. Moreover, the significance of DNA methylation in plant responses and adaptations to stresses such as drought or extreme temperature has been well documented. Because methylation changes are stable and heritable, the methylation pattern can prove to be a valuable marker for crop improvements. Thus, DNA methylation appears to be a crucial mechanism for plant developmental processes and protection from environmental stresses.

3.9 Further readings

Bancroft, Ian. 2001. "Duplicate and Diverge: The Evolution of Plant Genome Microstructure." *Trends in Genetics* 17 (2): 89–93. https://doi.org/10.1016/S0168-9525(00)02179-X.

Chiba, Yukako, and Pamela J. Green. 2009. "mRNA Degradation Machinery in Plants." *Journal of Plant Biology* 52 (2): 114–24. https://doi.org/10.1007/s12374-009-9021-2.

Feschotte, Cédric, and Ellen J. Pritham. 2007. "DNA Transposons and the Evolution of Eukaryotic Genomes." *Annual Review of Genetics* 41 (1): 331–68. https://doi.org/10.1146/annurev.genet.40.110405.090448.

Hames, B. D., and David M. Glover, eds. 1988. *Transcription and Splicing*. Frontiers in Molecular Biology. Oxford, England; Washington, DC: IRL Press.

Latchman, David S. 2002. *Gene Regulation: A Eukaryotic Perspective*. 4th ed. Cheltenham: Nelson Thornes.

Leaver, C. 1980. *Genome Organization and Expression in Plants*. Nato Asi Subseries A: Ser, v. 29. New York, NY: Springer.

Maniatis, Tom, and Robin Reed. 2002. "An Extensive Network of Coupling among Gene Expression Machines." *Nature* 416 (6880): 499–506. https://doi.org/10.1038/416499a.

Mukhopadhyay, Jigeesha, and Georg Hausner. 2021. "Organellar Introns in Fungi, Algae, and Plants." *Cells* 10 (8): 2001. https://doi.org/10.3390/cells10082001.

Ohyama, Kanji. 1996. "Chloroplast and Mitochondrial Genomes from a Liverwort, *Marchantia Polymorpha* —Gene Organization and Molecular Evolution—." *Bioscience, Biotechnology, and Biochemistry* 60 (1): 16–24. https://doi.org/10.1271/bbb.60.16.

Shi, Lan-Xin, and Steven M. Theg. 2013. "The Chloroplast Protein Import System: From Algae to Trees." *Biochimica et Biophysica Acta (BBA) - Molecular Cell Research* 1833 (2): 314–31. https://doi.org/10.1016/j.bbamcr.2012.10.002.

Yaish, Mahmoud W. 2013. "DNA Methylation-Associated Epigenetic Changes in Stress Tolerance of Plants." In *Molecular Stress Physiology of Plants*, edited by Gyana Ranjan Rout and Anath Bandhu Das, 427–40. India: Springer India. https://doi.org/10.1007/978-81-322-0807-5_17.

Chapter 4

Genetic Engineering Fundamentals

Genetic engineering (also called recombinant DNA technology and molecular biotechnology) uses a range of techniques for isolating specific genes, transferring them to other organisms and ensuring that they are properly expressed. To do this with bacteria (none of which have introns or inserted DNA), it is first necessary to cut a specific fragment of bacterial DNA with a restriction endonuclease, join it to a suitable (plasmid) vector with a ligase enzyme and then introduce it into a suitable host along with appropriate regulatory sequences. When starting with a eukaryote (generally an animal or plant) it is necessary to first isolate mRNA which is then converted into a double stranded DNA copy (i.e., cDNA). In that case, the cDNA is cloned onto an expression vector before the desired activity is detected. In addition, using the polymerase chain reaction (PCR) the number of specific pieces of DNA may be amplified a million times or more as part of its detection, isolation, and manipulation. Another, more recent, technique (CRISPR/Cas) enables scientists to modify the chromosomal DNA of any organism by adding to the existing DNA, removing portions of the existing DNA, or modifying the existing DNA.

4.1 Recombinant DNA technology

In the early 1970's scientists Stanley Cohen and Herbert Boyer made groundbreaking advancements to the study of biological processes at the molecular level. Before that period, all products made by living cells or cell components were naturally occurring and to produce large amounts of any protein or other compound, scientists had to develop methods for the large-scale cultivation of cells producing the desired protein or compound. The advent of recombinant DNA (rDNA) technology allowed scientists to insert genes from other organisms into bacterial (usually *Escherichia coli*) cells, express the foreign protein and grow large amounts of the *E. coli* cells enabling the production of specific proteins. Key to the development of recombinant DNA technology was the discovery of enzymes that could cut DNA at specific sites, i.e., restriction endonucleases, by Nobel laureates Werner Arber, Daniel Nathans, and Hamilton

O. Smith. The existence of restriction enzymes gave scientists the tools to not only analyze DNA more effectively, but also the ability to reproducibly cut DNA at specific locations so that it could later be joined together with other pieces of DNA to create new recombinant DNA fragments. This procedure, typically referred to as gene cloning, helped scientists (once a target DNA fragment was inserted into a cloning vector and the recombinant vector was introduced into *E. coli*) to make thousands of copies of the same DNA molecule, thus generating sufficient DNA for various kinds of experimentation and analysis. The development of rDNA technology has led to many new and innovative ways for isolation, analysis and manipulation of DNA and RNA.

Restriction enzymes, also known as restriction endonucleases, are crucial to many rDNA techniques. These enzymes, isolated from bacteria, recognize and cut (or digest) DNA at specific palindromic sequences in the DNA. Restriction enzymes exist in bacteria as a means of protection against destruction by DNA bacteriophages. A palindrome is a sequence of nucleotides that is read the same forward and backward on complementary strands (Fig. 4.1).

5'-----GAATTC-----3'
3'-----CTTAAG-----5'

Fig. 4.1 An example of a palindromic sequence in DNA

Several hundred different restriction enzymes are known with many of them currently commercially available; these enzymes typically recognize four, six or eight base sequences. A few frequently used restriction enzymes in rDNA technology are listed in Table 4.1. Note that the nomenclature for naming restriction enzymes includes the first letter of the bacterial genus and the first two letters of the bacterial species.

Table 4.1 Selected restriction enzymes and their bacterial sources

Enzyme	Bacterial source
EcoRI	*Escherichia coli*
BamHI	*Bacillus amyloliquefaciens*
HindIII	*Haemophilus influenza*
XbaI	*Xanthomonas badrii*
SalI	*Streptomyces albus*
PstI	*Providencia stuartii*

In a DNA cloning experiment, the choice of restriction enzyme depends on the specific requirements of that experiment. Four-base cutters are often used for generating smaller DNA fragments suitable for certain types of analysis, while eight-base cutters are used when many fewer and much larger fragments are desired. In nature, most restriction enzymes are six base cutters. Large genomes are usually cut with enzymes that produce long DNA fragments which can be easier to analyze and manipulate. Some enzymes cut straight across the DNA at the same position resulting in so-called blunt ends, others cut DNA within or near their recognition sites producing short single stranded overhangs, also called sticky ends. These sticky ends are complementary and are therefore able to base-pair with other DNA fragments with the same sticky ends (Fig. 4.2).

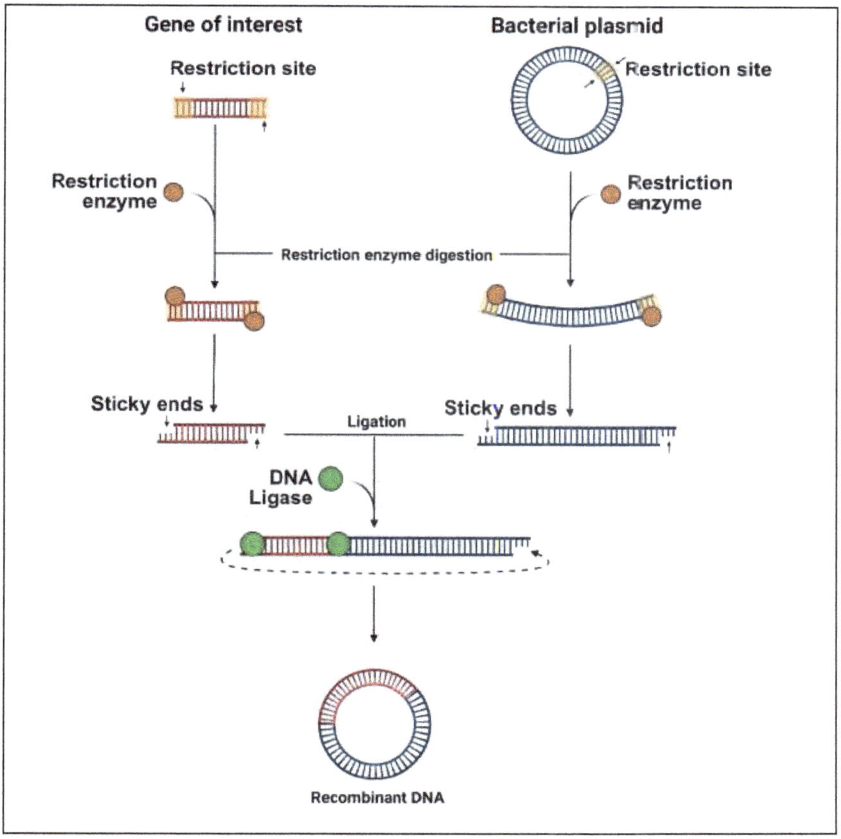

Fig. 4.2 Schematic of making recombinant DNA using a restriction enzyme and a DNA ligase

When cut by the same restriction enzyme, the resulting DNA fragments have identical sets of sticky ends and may be joined together by using an enzyme called DNA ligase (which is commercially available). DNA ligase facilitates the formation of phosphodiester bonds, allowing the joining of DNA fragments either from different molecules or the ends of the same molecule. ATP or NAD^+ is used to provide the energy needed for ligase to work. The reaction is often carried out at low temperatures (10°C) to improve ligation efficiency when dealing with sticky ends. However, the reaction is much slower at low temperatures, thus longer incubation times are needed to compensate for the low activity of ligase at this temperature. Once the phosphodiester bond is formed, the backbone of the DNA is fully restored, creating an rDNA molecule (Fig. 4.3).

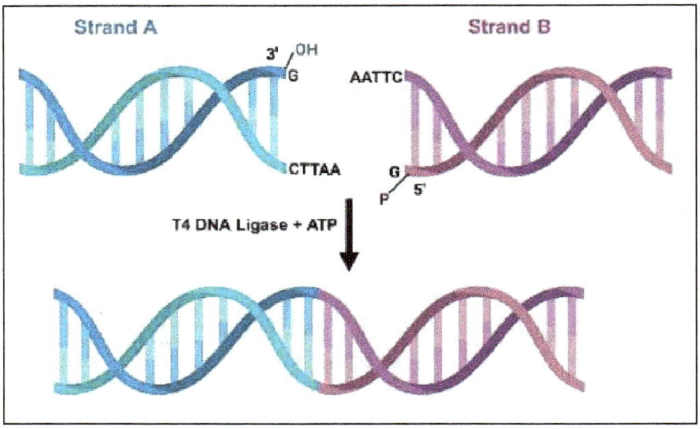

Fig. 4.3 The DNA products with sticky ends (held together by weak hydrogen bonds) resulting from restriction enzyme digestion may be fused together by the action of T4 DNA ligase. The DNA ligase forms a covalent bond between the 5' phosphate at the end of one strand and the 3' hydroxyl of the adjacent strand producing a continuous recombinant DNA molecule

The ligated recombinant DNA is propagated by inserting it into a host cell by using a special type of DNA molecule termed as a cloning vector. Many types of DNA molecules can serve as vectors, provided they have certain features. For instance, one of the key features of a good vector is self-replication; once in a host cell, a vector must be capable of replicating to a reproducible number of copies per host cell. Any foreign DNA that is inserted in the vector will be replicated in the process. The size of the vector is another crucial factor in choosing a vector, as different vectors have varying capacities for inserting foreign DNA. In general, larger vectors provide the necessary capacity to study extensive genetic information but they are slightly challenging to handle. On the other hand, smaller vectors typically have one or more unique restriction enzyme sites and are usually easy to manipulate which increases the efficiency of uptake by bacteria, a process termed as transformation. Stability is another important property of vectors. Many bacterial plasmids (extra chromosomal DNA) are small and circular in shape (increases transformation efficiency) and so are excellent potential vectors. Notice in the figure that the DNA of a plasmid is typically circular (Fig. 4.4).

Vectors for cloning DNA fragments are engineered in such a way that they facilitate easy linking of foreign DNA and selection of recombinants from non-recombinants. Plasmid pBR322 is an example of one of the first artificially modified cloning vectors. Named after its developers Francisco Bolivar and Raymond Rodriguez, pBR322 (where the p stands for plasmid), has a number of crucial elements that make it highly desirable – a bacterial origin of replication (ORI) to efficiently propagate within the bacterial host cell, several useful restriction enzyme site(s) where foreign DNA may be inserted and genes coding for resistance to the antibiotic's - ampicillin and tetracycline (Fig. 4.4). In this case, the two antibiotic resistance genes each possess unique cloning sites for the insertion of foreign DNA fragments and are required for selection of transformed cells. For example, insertion of the foreign DNA at the BamHI site inactivates the tetracycline resistance gene. When it is necessary to isolate cells containing the vector, cells are plated onto medium containing ampicillin. Colonies which grow on these plates must be derived from bacteria which contain the plasmid vector due to the expressed antibiotic. Selection of recombinant DNA molecule is achieved by plating the cells in the presence of tetracycline. Thus, colonies which grow on ampicillin but not on tetracycline must contain the recombinant plasmid. This indicates the relevance of the second antibiotic resistance gene in this vector.

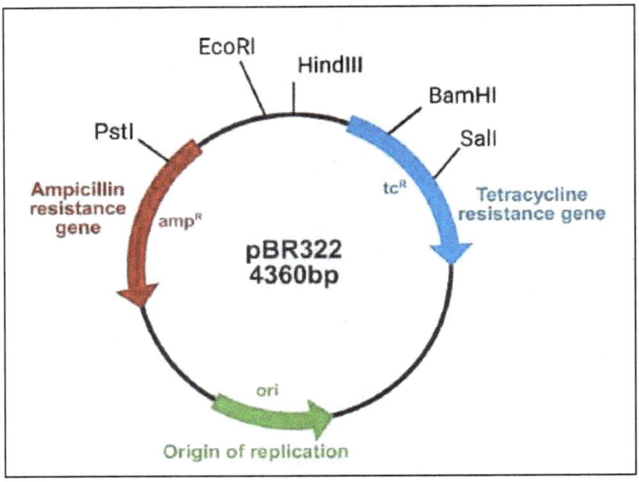

Fig. 4.4 Map and important features of plasmid pBR322. Two genes one encoding resistance to the antibiotic ampicillin (amp^R) and the other encoding resistance to the antibiotic tetracycline (tc^R) can serve as marker genes. All of the enzymes indicated (PstI, EcoRI, HindIII, BamHI, and SalI) cut the plasmid DNA only once and therefore can be used for the insertion of foreign DNA. An origin of replication (ori) allows the plasmid to be self-replicating within a bacterial cell

In some other plasmid vectors, for instance, pUC18 (developed at the University of California), the most popular restriction sites are concentrated into a region commonly referred to as multiple cloning site or MCS (Fig. 4.5).

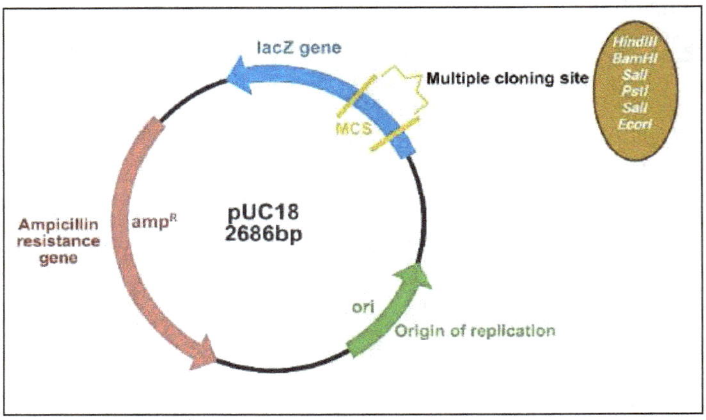

Fig. 4.5 Map of plasmid pUC18 showing important features including the fact that the multiple cloning site contains six different unique restriction enzyme sites

Moreover, since the MCS in pUC18 is located within the *lacZ* gene which encodes β -galactosidase it is possible to detect those plasmids which contain a DNA insert. The presence of a chromogenic substrate yields blue colored bacterial colonies following the substrate's cleavage by β -galactosidase, if they were derived from cells which contain the plasmid without an insert. However, if the *lacZ* gene is disrupted by the insertion of foreign DNA fragment, the colonies will appear white or colorless because β -galactosidase is no longer formed. Thus, a recombinant pUC18 plasmid may be readily detected on Petri plates by an easily identifiable color reaction. This elegant system, termed blue/white screening, named after the color of the bacterial colonies formed at the end of the screening procedure, allows quick identification of desired recombinant plasmids and has been utilized in subsequent commercially produced cloning vectors. In this regard, a typical workflow for recombinant DNA technology is shown in Fig. 4.6.

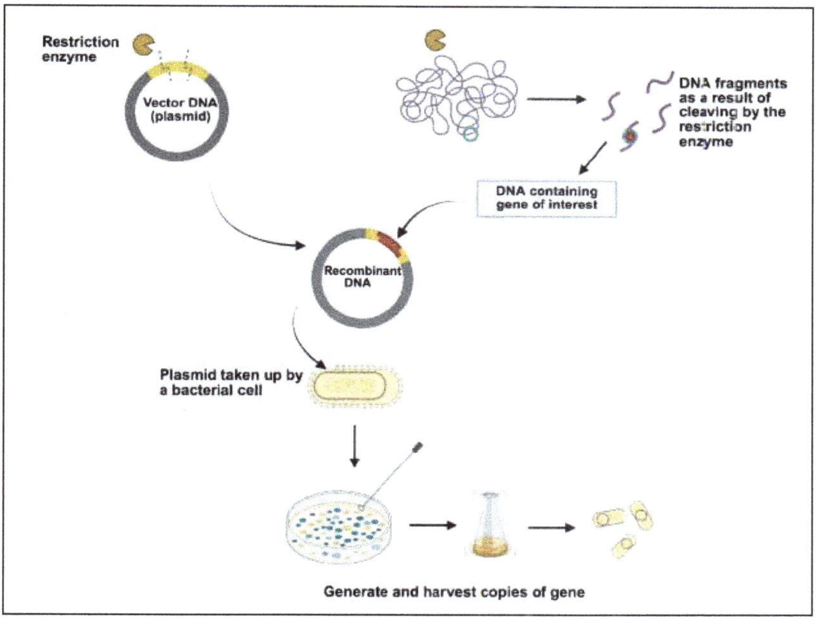

Fig. 4.6 Overview of the rDNA process

In many cases, it is desirable to study specific genes. The cDNA or complementary DNA method is frequently used method of obtaining eukaryotic genes. As we have seen earlier, eukaryotic genes generally consist of both short coding sequences, exons, and much longer non-coding sequences, introns; when the RNA transcript of such a gene is converted to mRNA, the introns are removed from the initial RNA transcript. Thus, when the aim is to clone genes of eukaryotic cells, it is preferable to start with a cDNA. This process is achieved by isolating the mRNA from a particular tissue or cells and then making a complementary DNA copy by using an enzyme called reverse transcriptase (discovered separately by Drs. Howard Temin and David Baltimore who shared the Nobel Prize for this work) and then to make double stranded (ds) DNA with DNA polymerase (Fig. 4.7) This synthesis is the reverse of the transcription (DNA to RNA) process. Each clone obtained in this way is called a cDNA clone, and could be inserted into an expression vector and would be available for genetic manipulation and modifications. The entire collection of clones derived from the original mRNA pool forms a cDNA library. However, it is important that the mRNA is of high quality and since mRNA constitutes only a small fraction (2-5%) of the total RNA within a eukaryotic cell so that isolating pure mRNA can be difficult. Another problem here is that sometimes the synthesized cDNA lacks the 5' end of the gene, probably because of random termination of reverse transcription. Nevertheless, this problem has been overcome by refined cDNA synthesis techniques.

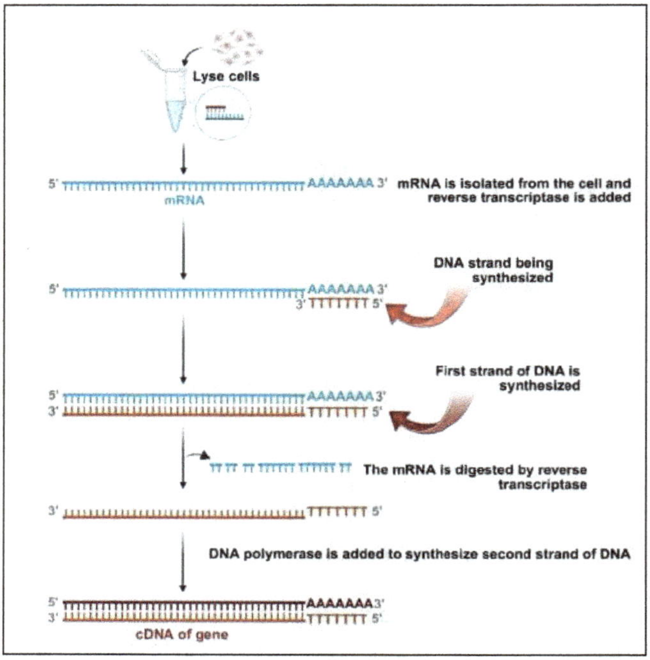

Fig. 4.7 A general scheme of making of complementary DNA (cDNA) for a gene from mRNA, using reverse transcriptase enzyme

4.2 Polymerase chain reaction

The polymerase chain reaction (PCR), developed by Dr. Kary Mullis in the 1980s, is the enzymatic amplification of small pieces of DNA to produce quantities that are large enough for analysis. One of the reasons for the widespread adoption of PCR is the simplicity, speed and efficiency of this reaction – it can make millions to billions of copies of a specific fragment of DNA in only a few hours, from even the smallest samples. PCR has rapidly become one of the key techniques used to analyze and clone DNA, and it has since revolutionized molecular biology, earning Kary Mullis the Nobel Prize in 1993.

For PCR, the starting material, usually termed template DNA, can come from a variety of sources and does not need to be highly purified. However, PCR does require prior knowledge of the DNA sequence which flanks the target DNA (the DNA fragment to be amplified). Using this information, two small oligonucleotide primers are chemically synthesized that are complementary to the flanking regions of the target DNA. The primers which are optimally 18-24 nucleotides long bind to their complementary sequence on the DNA samples. The primers are recognized by a thermostable version of the enzyme DNA polymerase (isolated from the thermophilic bacterial species *Thermus aquaticus*), which extends the primers in presence of other reagents including deoxyribonucleotides (dNTPs: dATP, dCTP, dGTP, and dTTP), which are the precursors to the nucleotide bases found in DNA, and a buffer appropriate for the enzyme to synthesize new complementary strands. Not all of the DNA fragment will be copied; only the region of DNA past where the primer binds (at the 3'end) will be amplified. The PCR reaction consists of three specific steps: (i) denaturation (ii) annealing and (iii) extension, as shown in Fig. 4.8.

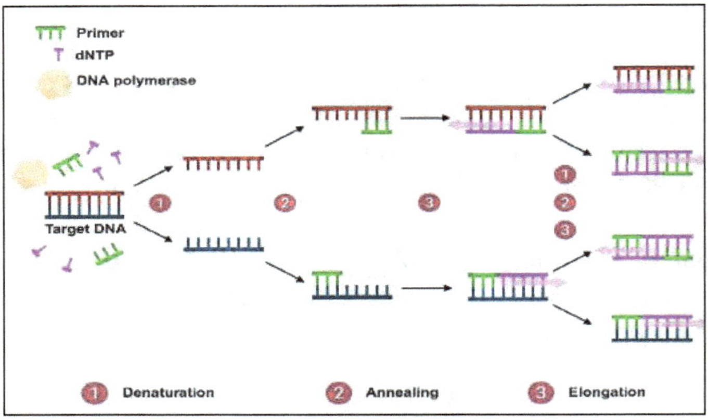

Fig. 4.8 Polymerase chain reaction (PCR). Each cycle consists of the three indicated steps and each cycle is repeated 30-40 times

These steps, termed cycles, are repeated 30-40 times in order to amplify the desired product a million to a billion times. In the first cycle, the double stranded template DNA is denatured by heating the reaction to 90-98°C. At this high temperature, the two strands separate, and the target DNA is made accessible to the primers. The reaction mixture is cooled to 40-60°C allowing the primers to bind to their complementary sites that flank the target DNA. Subsequently, the annealed (bound) primers serve as starting points for DNA synthesis, since they provide a free 3'-OH hydroxyl group necessary for DNA polymerase functioning. The duration and temperature of each step within a cycle may be modified to optimize the production of the desired amplicon and to prevent amplification of non-specific products. The DNA synthesis step is termed as extension and is carried out in the presence of *Taq* DNA polymerase and the essential deoxyribonucleotides. The strands synthesized in one cycle act as a template in the next cycle, resulting in an exponential increase in the amount of DNA produced. Twenty cycles, completed in just a few hours, will increase the amount of target DNA by more than a million times and after 30 cycles, it can reach over a billion copies. As a result, the process proceeds exponentially producing large quantities of DNA from a very limited amount of starting material making PCR a valuable tool in the core biosciences and biotechnology. The amplified DNA fragments may be cloned, sequenced or used as a diagnostic tool.

The entire PCR process is performed in a small plastic tube that is located within a thermal cycler which is an instrument that automates the temperature changes and incubation times for PCR (Fig. 4.9).

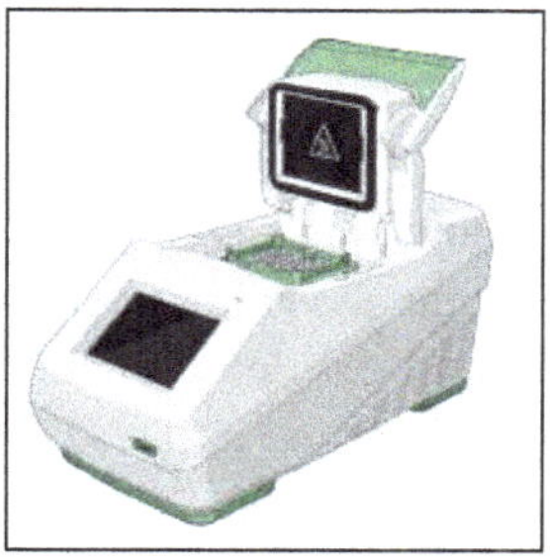

Fig. 4.9 A thermocycler

The DNA fragments amplified by PCR can be separated by a technique known as gel electrophoresis (Fig. 4.10). This technique separates DNA fragments according to their length i.e., their number of base pairs (bp). Gels are prepared using agarose (a long and linear carbohydrate polymer) dissolved in a buffer, warmed to dissolve the agarose, then poured into a form to harden. DNA fragments migrate through the gel towards the positive electrode under the influence of an electric field. Smaller fragments travel further (faster) than larger ones over time. The separated fragments can be visualized by staining the DNA with the dye ethidium bromide. This dye binds to DNA by insertion between stacked base pairs and exhibits a reddish-brown fluorescence when exposed to ultraviolet light. Alternative stains to ethidium bromide, such as SYBR® safe or SYBR® Green are also available. These have similar sensitivities to ethidium bromide but are less hazardous to use.

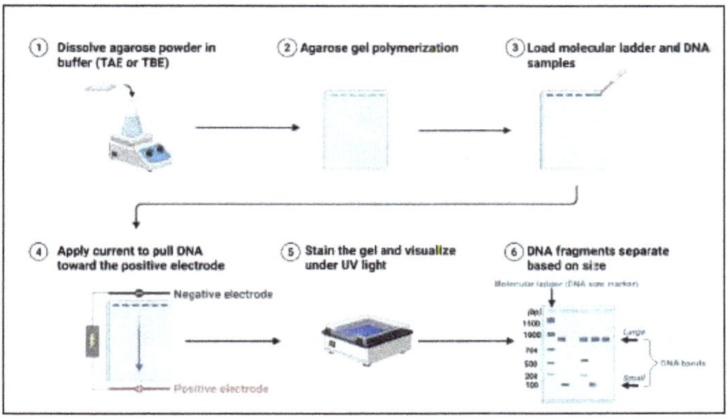

Fig. 4.10 Schematic overview of the process of agarose gel electrophoresis. Fragments of the same size stay together and migrate at the same rate through the gel matrix and appear as a single band

4.3 CRISPR/Cas

Although the discovery of restriction enzymes was a turning point that fueled the era of rDNA technology, the ability to introduce specific changes into whole genomes of microbes, plants and animals termed as genome editing marked a significant advancement beyond simple DNA cutting and pasting. To this end, several key developments were revealed in the 1980's about clustered regularly interspaced short palindromic repeats (CRISPR) sequences detected in bacteria and archaea. Originally thought to be involved in gene regulation, the function of these sequences as an adaptive defense against viral infections was apparent a decade later. Subsequent discoveries led to identification of the CRISPR-associated Cas genes which are well conserved across bacterial species. The system works by storing pieces of viral DNA in the form of non-repeating spacer sequences. Upon infection with the same virus, the bacterium can recognize and destroy it with the enzymes encoded by Cas genes. Among the many Cas proteins, Cas9 is responsible for cleaving DNA. Cas9 activity relies on CRISPR RNA (crRNA) and a trans-activating CRISPR RNA (tracrRNA). These two RNAs form the Cas9 protein-RNA complex known as a guide RNA (gRNA). The gRNA searches for a target site in the target DNA (in nature, often a bacteriophage), and after a successful complementary match, the Cas9 protein acts as an endonuclease and cuts the target DNA sequence, three bp away from the PAM (protospacer adjacent motif, a short sequence downstream of the target site) resulting in a double-stranded break in the DNA (Fig. 4.11). By modifying the 20-nucleotide spacer (variable) region of the guide RNA (gRNA), researchers can reprogram the CRISPR/Cas9 system to cleave any DNA sequence of interest — provided it is adjacent to an appropriate PAM sequence. These observations were the result of work published in 2012 by two women scientists, Drs. Jennifer Doudna and Emmanuelle Charpentier, who expanded the understanding of the molecular mechanisms that enable the natural CRISPR/Cas9 system to cut DNA. These two scientists received the Nobel Prize in Chemistry in 2020 for this work. Furthermore, they also demonstrated how to program the Cas9 protein with a synthetic guide RNA (gRNA) that could direct the Cas9 enzyme to cleave at specific DNA target sites (see Fig. 4.11). Thus, the technology has been adopted in targeted gene deletion, alteration and insertion in many plant species.

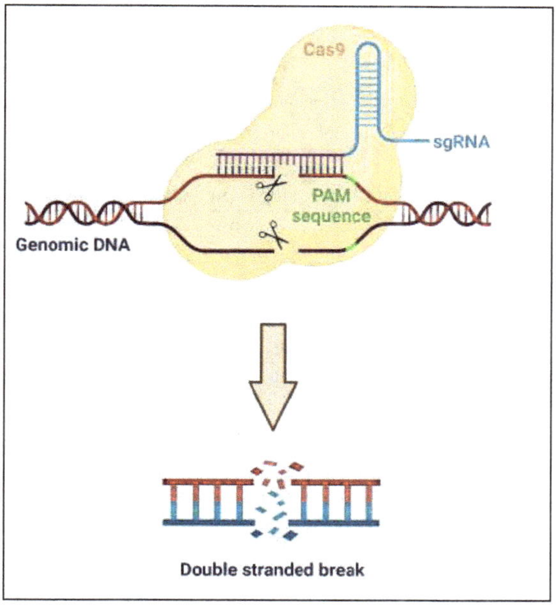

Fig. 4.11 Schematic of a CRISPR/Cas9 genome-editing system with a guide RNA (gRNA) targeting a specific DNA sequence adjacent to PAM; DNA cut sites are shown using scissors

The CRISPR/Cas9 technique enables scientists to create precise cuts at a range of target sites in most crop plants, and thus largely increases the speed of crop improvement. Since its discovery as a plant genome editing technique, the technology has been applied successfully in about 130 crops and model plants. The use of CRISPR/Cas in plant biotechnology is particularly exciting as this system is viewed as a technically simple means of altering a particular plant trait (Table 4.2).

Table 4.2 Some examples of commercialized CRISPR/Cas edited crops

Product	Trait	Status
Maize	High yield, waxy	Pre-commercial
Mushroom	Non-browning	Released, 2021
Tomato	High GABA	Released, 2021
Mustard greens	Reduced natural pungency	Awaiting approval
Lettuce	Non-browning	Pre-commercial
Soybean	High oleic acid High protein	Released, 2019 Released, 2024
Canola	Herbicide tolerance Increased Omega-3 fatty acid	Released, 2019 Pre-commercial
Rice	High yield, drought and salt tolerance	Released, 2025
Camelina	High oil	Pre-commercial

With CRISPR, scientists can do many things—like insert new genes to give an organism a new trait, delete parts of genes to generate a knock-out mutant that no longer encodes for the original protein, to study their function or remove unwanted traits, and modify existing genes to fix mutations or improve how they work or replace the existing gene with a foreign gene (Fig. 4.12).

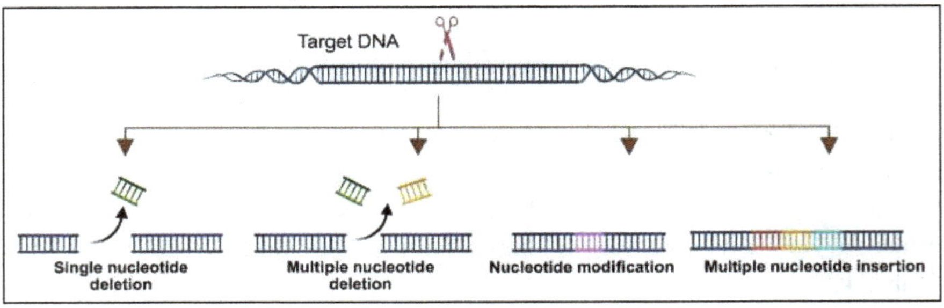

Fig. 4.12 Schematic showing plant DNA cut in a targeted manner by a CRISPR/Cas9 complex, represented by scissors. (a) single or (b) multiple nucleotide deletion, generating a knock-out mutant that no longer encodes the original protein (c) nucleotide modification, may alter the protein function (d) multiple nucleotide insertion may lead to altering the protein function or it may replace the existing gene with a foreign gene

A technical limitation of the CRISPR tool is that it does not always cut at the targeted site in the genome. Recognizing this problem, scientists have repurposed the CRISPR/Cas system to improve its targeting accuracy, sometimes by modifying the gRNA or tweaking the Cas protein. New methods are continuously being invented, for instance, by fusing a protein to a deactivated Cas9 nuclease (dCas9) researchers can directly change DNA bases (Fig. 4.13).

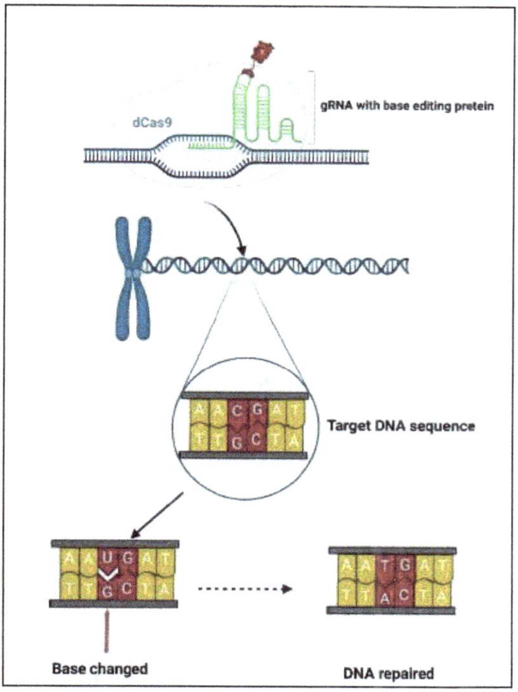

Fig. 4.13 An overview of base editing using CRISPR

4.4 Further readings

Anzalone, Andrew V., Luke W. Koblan, and David R. Liu. 2020. "Genome Editing with CRISPR–Cas Nucleases, Base Editors, Transposases and Prime Editors." *Nature Biotechnology* 38 (7): 824–44. https://doi.org/10.1038/s41587-020-0561-9.

Berger, Shelby L., and Alan R. Kimmel. 1987. *Guide to Molecular Cloning Techniques*. Methods in Enzymology, v. 152. Academic Press.

Bolivar, F., R. L. Rodriguez, P. J. Greene, M. C. Betlach, H. L. Heyneker, H. W. Boyer, J. H. Crosa, and S. Falkow. 1977. "Construction and

Characterization of New Cloning Vehicles. II. A Multipurpose Cloning System." *Gene* 2 (2): 95–113.

Cohen, S.N., Chang, A.C.Y., Boyer, H.W., and Helling, R.B. 1973. Construction of Biologically Functional Plasmids *In Vitro*. *Proc. Natl. Acad. Sci. USA*. 70 (11): 3240-3244.

Doudna, Jennifer A., and Emmanuelle Charpentier. 2014. "The New Frontier of Genome Engineering with CRISPR-Cas9." *Science* 346 (6213): 1258096. https://doi.org/10.1126/science.1258096.

Glick, Bernard R., and Cheryl L. Patten. 2022. *Molecular Biotechnology: Principles and Applications of Recombinant DNA*. Sixth edition. ASM Press.

Khehra, Nimrat, Inderbir S. Padda, and Cathi J. Swift. 2025. "Polymerase Chain Reaction (PCR)." In *StatPearls*. Treasure Island (FL): StatPearls Publishing. http://www.ncbi.nlm.nih.gov/books/NBK589663/.

Pixley, Kevin V., Jose B. Falck-Zepeda, Robert L. Paarlberg, et al. 2022. "Genome-Edited Crops for Improved Food Security of Smallholder Farmers." *Nature Genetics* 54 (4): 364–67. https://doi.org/10.1038/s41588-022-01046-7.

Roberts, Richard J. 2005. "How Restriction Enzymes Became the Workhorses of Molecular Biology." *Proceedings of the National Academy of Sciences* 102 (17): 5905–8. https://doi.org/10.1073/pnas.0500923102.

Saiki, Randall K., David H. Gelfand, Susanne Stoffel, Stephen J. Scharf, Russell Higuchi, Glenn T. Horn, Kary B. Mullis, and Henry A. Erlich. 1988. "Primer-Directed Enzymatic Amplification of DNA with a Thermostable DNA Polymerase." *Science* 239 (4839): 487–91. https://doi.org/10.1126/science.2448875.

Tsai, Shengdar Q., and J. Keith Joung. 2016. "Defining and Improving the Genome-Wide Specificities of CRISPR–Cas9 Nucleases." *Nature Reviews Genetics* 17 (5): 300–312. https://doi.org/10.1038/nrg.2016.28.

Tuncel, Aytug, Changtian Pan, Thorben Sprink, Ralf Wilhelm, Rodolphe Barrangou, Li Li, Patrick M. Shih, et al. 2023. "Genome-Edited Foods." *Nature Reviews Bioengineering* 1 (11): 799–816. https://doi.org/10.1038/s44222-023-00115-8.

Chapter 5

Genetic Engineering of Plants

About a decade after the emergence of rDNA technology, genes from the bacterium *Agrobacterium tumefaciens* were used to carry foreign genes into plants. The vast majority of plants may be genetically transformed to express foreign genes either using the Ti plasmid from the bacterium *Agrobacterium tumefaciens* or by introducing the foreign DNA by microprojectile bombardment. The foreign gene(s) may be directed to the nuclear or the plastid DNA, its expression may be regulated using appropriate transcriptional signals, the carbohydrates attached to foreign proteins may be manipulated to resemble glycosylation patterns normally found in plants, and the foreign genes may be introduced in the absence of problematic marker genes. Moreover, using the CRISPR/Cas system, existing plant genes may be modified *in situ* resulting in variants of the original plant genome.

5.1 Ti plasmid and derived vectors

*Agrobacterium tumefaciens** (synonym: *Rhizobium radiobacter*) is a soil borne plant pathogen that causes crown gall disease in plants. Crown gall disease, often referred to as a cancerous growth in plants due to its uncontrolled cell division and tumor formation, has a long and fascinating history. What makes this disease particularly intriguing is that *A. tumefaciens* manipulates its host by transferring a portion of its genetic information to the plant, inducing tumor formation, and instructing the resulting tumors to make nutrients to support its growth. This manipulation effectively turns the plant into a biological factory, producing both tumors and specialized nutrients that benefit the invasive bacterium.

Several *Agrobacterium* species are known for their ability to infect a wide range of plant species. However, successful infection and tumor formation occur only when the bacterium encounters a wound on a susceptible plant (Fig. 5.1). A considerable amount of work has been done to understand the host range and the specific conditions required for plant infection by *Agrobacteria*. Research about the infection process, has led to the understanding that as part of its natural parasitic life cycle, *A. tumefaciens* transforms plant cells by transferring a segment of its plasmid DNA, called T-DNA (or transfer DNA), into the plant chromosomal DNA (Fig. 5.1). When this happens, the plant cell becomes genetically modified leading to the formation of crown gall tumors. Inside the plant cell, the T-DNA is replicated and expressed along with the plant's own DNA, causing the plant cells to divide and grow abnormally. This natural mechanism of gene transfer has been exploited by scientists to insert foreign genes within the T-DNA, so that they could be transferred to the chromosomal DNA of the host plant.

In order to be able to regenerate plants from cells transformed with T-DNA in culture, scientists modified the Ti plasmid (discussed below) by removing the tumor causing genes and replacing them with the desired genes.

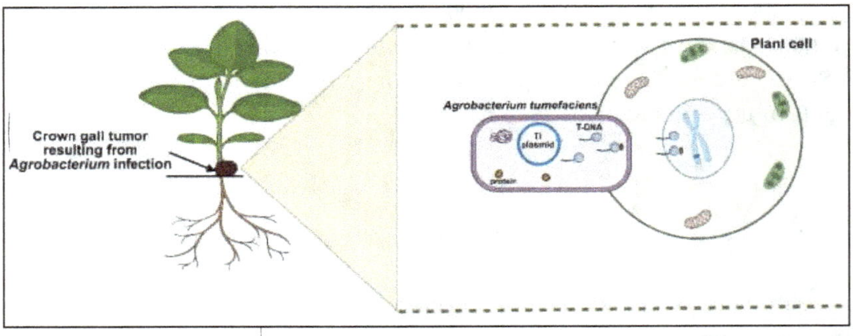

Fig. 5.1 An overview of *Agrobacterium*-mediated T-DNA transfer into a plant genome. (**Agrobacterium tumefaciens* (now formally classified as *Rhizobium radiobacter*) is the updated taxonomic designation; however, for clarity and consistency with common usage in plant biotechnology, we will refer to it by its former name, *Agrobacterium tumefaciens*, throughout this text)

The T-DNA is part of a much larger tumor-inducing (Ti) plasmid and interestingly, only this relatively small T-DNA fragment is transferred into the plant, while the rest of the Ti plasmid remains in the bacterium. The formation of plant tumors is initiated by the wound-induced release of signaling compounds that trigger a series of events that lead to the integration of the T-DNA into the nucleus of the plant cell.

A critical portion of the Ti plasmid which controls the transfer process is the virulence (*vir*) region. It is activated by plant phenolic molecules such as acetosyringone and α-hydroxyacetosyringone (Fig. 5.2) and comprises approximately 30 genes organized into several operons that are regulated together in response to the plant signal molecules. The specific functions of each of these operons are listed in Table 5.1.

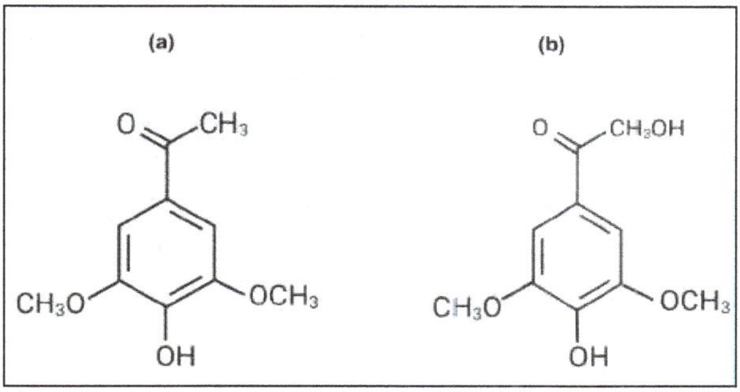

Fig. 5.2 Inducers of Ti plasmid *vir* genes (a) acetosyringone (AS) (b) α-hydroxyacetosyringone (OH-AS)

Table 5.1 *Vir* operon types and functions in *Agrobacterium*-mediated transformation

***Vir* operon**	**Specific function**
VirA	Detects plant signals (e.g., acetosyringone) and activates *VirG*
VirG	Binds to all *vir* gene promoters and activates their transcription
VirB	Creates a pilus that transfers the T-DNA/protein complex into plant cells
VirC	Enhances T-DNA processing efficiency and stabilization
VirD	Excises the T-DNA at the 25 bp left border & right border repeats, then physically links the T-DNA to the *VirD2* protein and guides the T-DNA complex from *Agrobacterium* to the plant nucleus
VirE	Protects the T-DNA from degradation and helps in nuclear import
VirF (not always present)	Degrades plant proteins that could inhibit T-DNA integration

The T-DNA region of the Ti plasmid encodes genes that direct production of the plant hormones auxin (*iaaM* or *iaaH*) and cytokinin (*ipt*) whose overproduction controls the tumor morphology. It also encodes enzymes needed for opine synthesis, which are compounds that are important to maintain the tumorous state in the infected cells. Opines are unusual condensation products of either an amino acid and a keto acid (e.g., nopaline) or an amino acid and a sugar (e.g., octopine), and they are synthesized by the tumorous cells and not by normal plant cells. Moreover, opines are subsequently utilized by *Agrobacterium* as a carbon and energy source. Some opines correlate with variations in auxin/cytokinin ratio influencing the tumor phenotype whether tumors are shoot-like, root-like, or callus-like. *Agrobacterium* strains and their Ti plasmids are distinguished based on the particular opines that are made in the tumors that they induce and that the strain can catabolize. Genes required for opine catabolism (e.g., *noc* for nopaline catabolism) are present on the Ti plasmid outside of the T-DNA region (Fig. 5.3). The synthesis of opines by the infected tissues activates expression of opine catabolic genes as well as genes that regulate the replication and conjugal transfer of the Ti plasmid.

A repeated sequence of 24 base-pairs (bp) is present at each end of the T-DNA, known as left and right borders. These borders play a crucial role in the transfer of DNA from the bacterium into plant genome. In fact, any gene inserted between the borders will be transferred to the plant genome, allowing scientists to introduce foreign genes not involved in tumorigenesis, such as those encoding beneficial traits, into plant genomes.

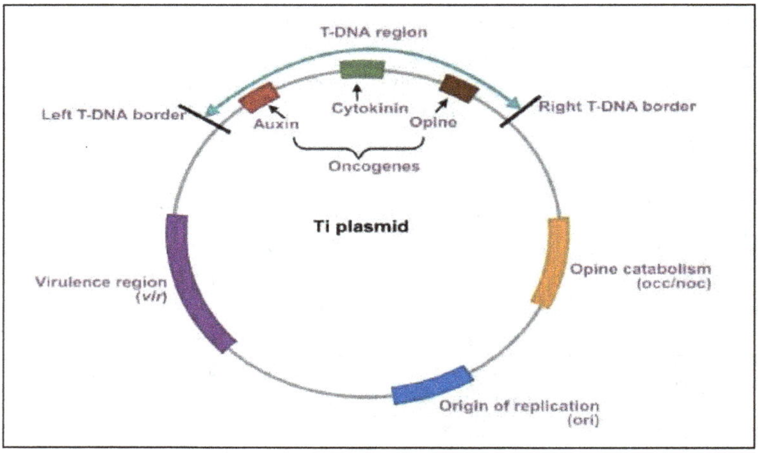

Fig. 5.3 Features of the circular nopaline pTiC58 (Ti plasmid from *A. tumefaciens* strain C58); The virulence (*vir*) region controls the transfer of the T-DNA. Highlighted portions of the Ti plasmid include the *vir* region, T-DNA borders, *noc* (opine catabolism gene), and conjugative transfer (*tra*) loci together with the plasmid's origin of DNA replication

A. tumefaciens is considered an ideal vector for plant transformation. However, a drawback of this system is its inability to control the T-DNA integration site as insertion occurs randomly within the plant genome. This may lead to gene silencing and unpredictable outcomes. Other limitations include the large size of the Ti plasmid (approx. 200 kbp) making it difficult to clone foreign gene sequences within the plasmid. In addition, the presence of auxin and cytokinin biosynthesis genes in the T-DNA induces uncontrolled cell division, leading to crown gall tumors, hindering the ability to regenerate whole plants from the transformed cells. Therefore, these genes must be removed. Additionally, the opine synthesis gene is not useful to transgenic plants as it diverts plant resources away from growth and productivity, potentially reducing plant yield. Thus, removing this gene is necessary for optimizing plant performance. Lastly, because the Ti plasmid does not replicate in *E. coli,* researchers cannot easily amplify or maintain modified versions of the plasmid in bacterial cultures.

Improvements of Ti plasmids have addressed some of the above-mentioned challenges. In general, the following features are essential for the construction of an efficient Ti-vector system:

i. An origin of replication (Ori) – ensures that the plasmid can be maintained and replicated in the host bacteria, typically *E. coli* for cloning.
ii. Plant and bacterial selectable marker genes – to distinguish between transformed and non-transformed plant tissues. For example, the neomycin phosphotransferase gene (NPTII) provides resistance to the antibiotic kanamycin, and the hygromycin phosphotransferase gene (HPT) confers resistance to the antibiotic hygromycin. Transformed tissues contain selectable markers and survive on selective media containing antibiotics, while non-transformed cells, which do not have an antibiotic resistance gene, are killed on the antibiotic media.
iii. Right and left T-DNA border sequence- the right border (RB) is the important site where T-DNA transfer is initiated, while the left border (LB) ensures complete transfer.
iv. Multiple cloning site (MCS) – A short DNA sequence with multiple restriction enzyme sites allows for the easy insertion of foreign genes. This flexibility enables researchers to insert genes of interest while maintaining compatibility with standard cloning techniques.

Thus, the following two types of Ti-plasmid vector systems have been developed.

Co-integrative vector system - A co-integrative vector is a type of Ti-derived vector in which the majority of the T-DNA between the border repeats has been replaced by a defined sequence of DNA. A foreign DNA sequence is cloned into an intermediate vector stable in *E. coli*, which contains selectable markers (for example, antibiotic resistance), and a sequence homologous to that present between the border repeats of the T-DNA of the Ti plasmid. This homologous region allows the intermediate vector to integrate into the Ti plasmid. The T-DNA region of the Ti plasmid is removed and cloned into a small *E. coli* plasmid, such as pBR322. The foreign gene is cloned within the pBR322, and it can be co-integrated into the disarmed Ti plasmid (forming a co-integrative unit) by homologous recombination (Fig. 5.4). Due to a lack of the appropriate origins of replication the intermediate vector is unable to be stably maintained within *Agrobacterium*, and the foreign DNA sequence is integrated into the co-integrative vector by homologous recombination. In this state, the modified T-DNA plasmid, which already carries the virulence genes that will bring about the transfer and integration of the modified T-DNA carrying the target genetic material into the plant genome.

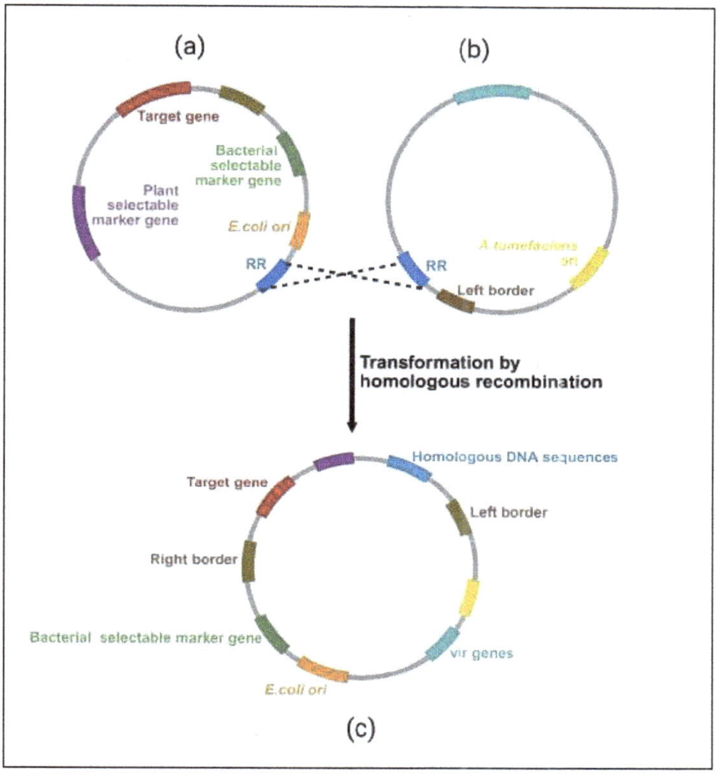

Fig. 5.4 (a) The cointegrate cloning vector contains an *E. coli ori*, so it cannot replicate independently in *A. tumefaciens.* It carries a selectable marker that works either in *E. coli* or *A. tumefaciens*, a T-DNA right border, a plant selectable marker, a target gene, and a region homologous to a segment of the disarmed Ti plasmid (b) the disarmed Ti plasmid contains the T-DNA left border, *vir* region, and *A. tumefaciens ori* (c) the plasmid, formed after recombination between plasmids (a) and (b) contains the complete T-DNA region including the target gene and the plant selectable marker flanked by left and right border sequences, along with the *vir* region necessary for T-DNA transfer

Binary vector system: In binary vector systems, the disarmed, helper Ti plasmid (which lacks a functional T-DNA) is partnered with a second plasmid carrying the T-DNA left and right border sequences. Within the border repeats are the appropriate cloning sites for insertion of foreign DNA, as well as a selectable marker that functions in plants. Because the wild-type Ti plasmid is very large it provides the T-DNA transfer functions in trans to the binary vector plasmid, which is small and easy to handle. The binary vector is constructed in *E. coli,* carrying an origin of replication (*ori*) and an antibiotic resistance marker for selection in bacterial culture, and is subsequently introduced into *A. tumefaciens*, which harbors the plasmid with a functional *vir* region, via conjugation or transformation. The T-DNA region from the binary vector is transferred to the plant by expression of the *vir* genes in the Ti helper plasmid. A typical binary cloning vector is shown in Fig. 5.5.

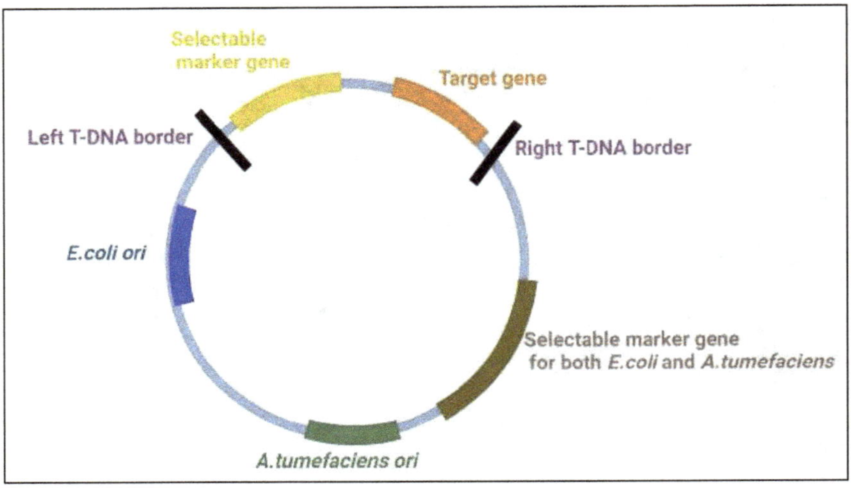

Fig. 5.5 Genetic organization of a binary vector indicating origins of replication (*ori*) for both *E. coli* and *A. tumefaciens*, a selectable marker gene functional in either *E. coli* or *A. tumefaciens*, a target gene and a plant selectable marker gene, both inserted between the left and right borders of the T-DNA to enable transfer and selection in transformed plant cells

Some commonly used methods for the delivery of gene constructs in plant cells via *Agrobacterium* are illustrated in Fig. 5.6.

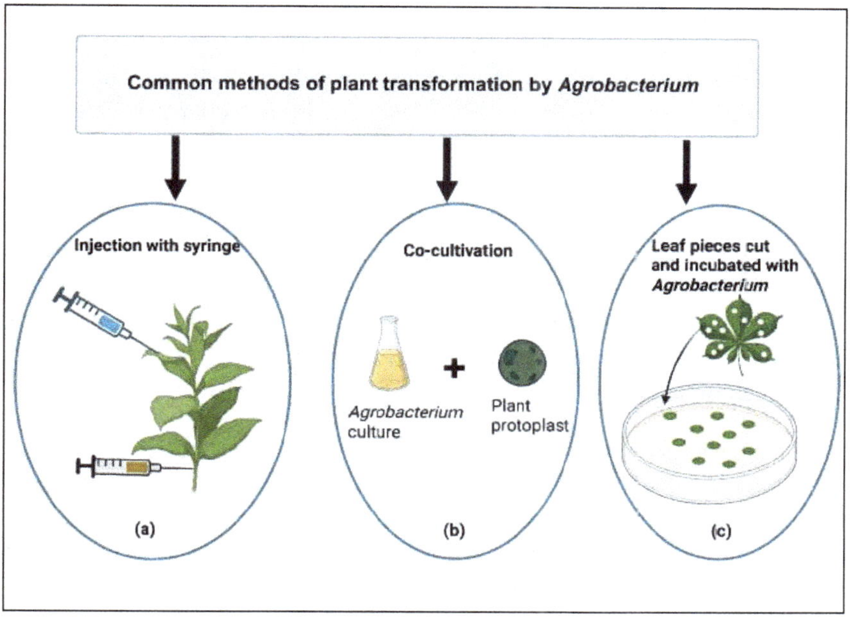

Fig. 5.6 (a) The stem inoculation method, in which a syringe with an injection needle is used to create a wound while simultaneously introducing a culture of *Agrobacterium* into the plant tissue (b) Co-cultivation of isolated plant protoplasts with an *Agrobacterium* culture. Protoplasts are plant cells with the cell wall removed (c) Leaf disc method, in which small sections of a mature leaf are excised using a punch and then incubated with an *Agrobacterium* culture to facilitate transformation. Other procedures involve developing a tissue culture system to generate transgenic plants from transformed cells

Agrobacterium-mediated plant transformation has been used widely but is limited by the host range of this bacterium. Monocots, including agronomically important crops such as rice, wheat, corn, soybean and barley, are recalcitrant to this type of transformation. A key requirement for transient expression and/or stable integration of DNA is the entry of T-DNA into the plant cell nucleus. This relies on the interactions of the transferred DNA with several bacterial proteins, and any inefficiency in this process can hinder the success of the technique. Also, if the objective is to study transient expression in plant cells and tissues, for which rapid evaluation is key, *Agrobacterium* inoculation may not be ideal. Typically, one has to start the *A. tumefaciens* culture 24 hours before inoculation, followed by 2-3 days of co-cultivating the bacterial cells with plant tissue and observation of transient expression. Some of these problems might be circumvented by using more direct methods of gene transfer (see the next section).

5.2 Microprojectile bombardment

It is possible to deliver gene constructs into living cells via high velocity microprojectiles which are tiny particles less than 4 µm in diameter. This method is known as microprojectile bombardment, and it is a major alternative to *Agrobacterium*-mediated gene transfer. Also referred to as biolistics, a term coined in the 1980's by Dr. John Sanford, there are several fundamental advantages of this technique over *Agrobacterium* transformation. It can be used with a wide variety of plant species, and tissue types including cell suspensions, callus cultures, meristematic tissues, pollen grains, embryos and more. Additionally, integration of foreign DNA is feasible not only in the nucleus, but also in other organelles, in particular chloroplasts.

The microprojectile technique involves coating microprojectiles of gold or tungsten with foreign DNA, placing them on a support, and accelerating them to high speed (usually 300-600 meters/sec) to the sample to be transformed, which is placed in an evacuated chamber. A pressurized gas, usually helium provides the force for the bombardment of the DNA-coated metal particles (Fig. 5.7a). After the bombardment, the foreign genetic material is released from the microprojectile and may be expressed transiently or it may become stably integrated into the chromosome of the plant cell. Sometimes, it is practical to use a hand-held device (Fig. 5.7b), which expands the choice of potential target material. Whereas, in the non-portable device the target tissue has to be placed in a chamber of limited size.

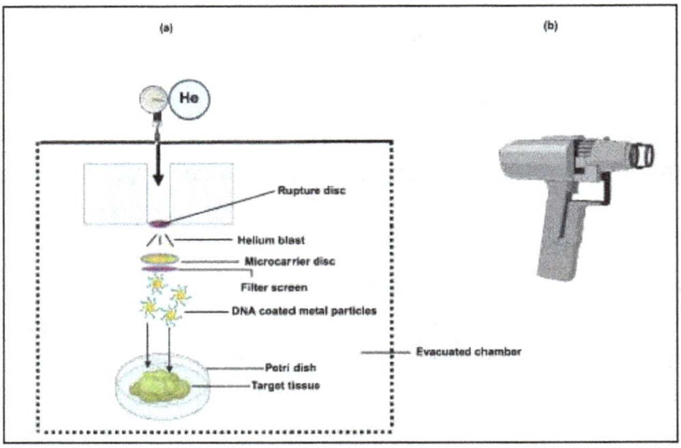

Fig. 5.7 (a) Schematic representation of a typical particle bombardment set up. It consists of an evacuated chamber with a rupture disc, a microcarrier disc with coated DNA particles, a filter screen and the target tissue (e.g., callus). Following the release of helium under pressure, the

plastic rupture disc breaks and the released gas accelerates the microcarrier disc with the DNA-coated metal particles. The rupture disc is blocked by the stopping screen and the DNA-coated metal particles pass through the screen toward the target tissue (b) a hand-held device that acts in a similar manner

The biolistic process is summarized below (Fig. 5.8).

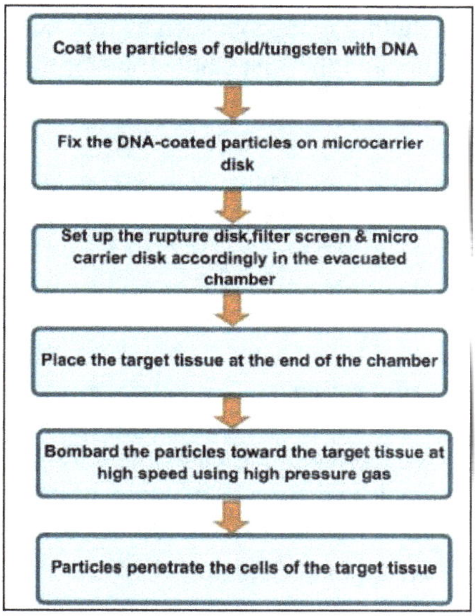

Fig. 5.8 General steps involved in the microprojectile bombardment process

Gene constructs for biolistics can be in the form of circular plasmids, as its purification from bacterial cultures provides a straightforward and efficient means to obtain sufficient DNA for bombardment experiments. It is common to use several plasmids with different combinations. Also, it is advantageous to the plant biotechnologist to evaluate the expression ability of a transgene before performing stable transformations.

There are some disadvantages to the biolistic process. For instance, the need for specialized equipment, which may limit its accessibility, and the gold particles can be costly. In order to obtain successful results, it is important to use an optimum amount of plasmid DNA. Too little plasmid DNA can result in a low level of plant cell transformation, while excessive plasmid DNA can be inhibitory to the process. Furthermore, the target tissue may get damaged due to the force of the impact of the microprojectiles. Thus, potential users may need to spend some time optimizing bombardment parameters such as chamber vacuum pressure, particle size and DNA: particle ratios, the gap between rupture disk and gold particles, the travel distance from gold particles to stopping screen, to their target species or tissues. Nevertheless, particle bombardment is often the method of choice due to its versatility and for its rapid selection process. At present, a number of successful microprojectile transformations have been achieved in a wide variety of crops including monocots such as rice, wheat and maize, barley, corn, onion, oats, sorghum, and the first stable transformation in a commercially important conifer species *(Picea glauca)* was achieved via embryogenic callus tissue as explant. Most recently, the use of particle bombardment for plant transformation has increased significantly, as it has become the preferred method of delivering the gene-editing elements (discussed later) into plant cells to effect targeted genome editing.

5.3 Chloroplast engineering

Each leaf cell may contain as many as 100 chloroplasts, and each chloroplast contains hundreds to thousands of copies of its genome. Consequently, when a foreign gene is inserted into the chloroplast, thousands of copies of the gene are subsequently present in each plant cell. Thus, exceptionally high levels of transgene expression may be achieved in the chloroplast. With chloroplast engineering, a human growth hormone somatotrophin protein accumulated to 7% of the total soluble protein in tobacco leaves, outperforming nuclear transgene expression by 300-fold. After that, researchers expanded into producing more specialized proteins, like antimicrobial peptides (e.g., plyGBS, a bacteria-phage lytic enzyme), which reached up to 70% of the of total leaf protein. The advantage of antimicrobial peptides over current antibiotics is that they are effective against drug-resistant microorganisms. This study points out the potential of chloroplast transformation to produce high levels of therapeutic proteins. Additionally, the chloroplast genome has been exploited to produce polyhydroxybutyrate (PHB), a renewable bioplastic. PHB synthesized in tobacco chloroplasts made up more than 18% of the plant's dry weight. A notable advantage of chloroplast transgenic plants (or transplastomic plants) over nuclear transgenic plants is that in most angiosperm species, the chloroplast DNA is inherited uniparentally, typically from the maternal (female) parent. This helps control the spread of genetically modified traits via pollen, reducing the risk of transgene escape to wild relatives on neighboring fields, which is a big advantage for subsequent public acceptance and regulatory approval. Clearly, there are considerable commercial opportunities in this area of plant biotechnology.

Genetic engineering of chloroplasts is based on homologous recombination. A key requirement for a vector used to integrate a foreign DNA into the chloroplast genome is that the target gene sequences must be flanked by chloroplast DNA sequences, a promoter with a 5' untranslated region (UTR) and a ribosome binding site. The 3' UTR includes a stem-loop structure which enhances transcript stability. An example of a chloroplast transformation vector is shown in Fig. 5.9. Homologous recombination can occur between chloroplast DNA sequences on the vector and the chloroplast genome. Therefore, the transgene must be flanked by homologous chloroplast DNA sequences on both sides to ensure stable integration of the transgene into the chloroplast genome.

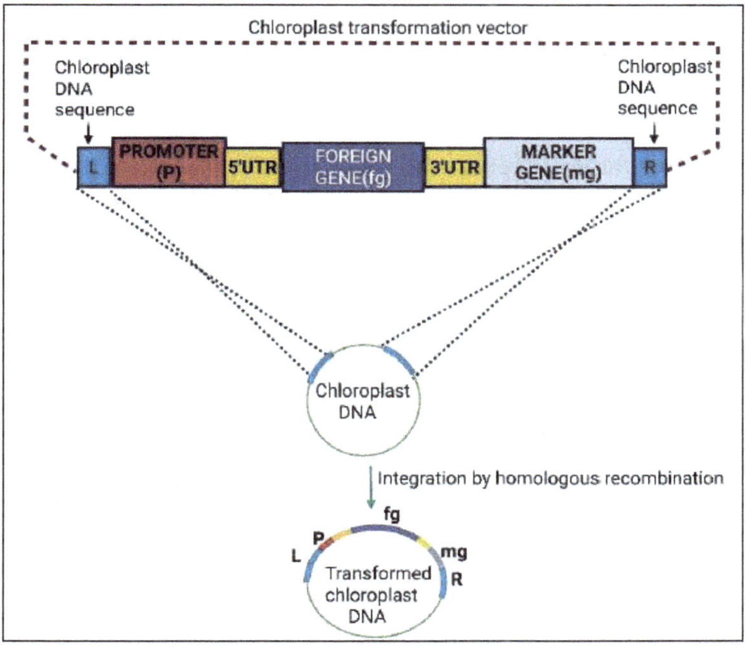

Fig. 5.9 A genetic map of a chloroplast-expression vector used for integrating a foreign gene (fg) and a marker gene (mg) into the native plastid genome. The fg and mg integrate into the plastid DNA through two homologous recombination events (indicated by dotted lines). The chloroplast flanking sequences, promoter, regulatory elements of foreign gene are shown- 5'-UTR, 5' untranslated region (to enhance ribosome binding), 3'-UTR, 3' untranslated region (to enhance transcript stability). The ribosome binding sites of both genes are not shown

Moreover, since chloroplast transcription and translation are prokaryote-like, vectors can be designed to introduce multiple genes (cistrons) in a single operon transformed into the chloroplast genome. This is especially useful to introduce biosynthetic pathways into plants, however one problem with this system is that it may result in poor gene expression of at least some transgenes. However, it is possible to use intercistronic expression elements or IEEs, which are small sequence elements that can be inserted between genes to enhance expression of all cistrons on the synthetic operon (Fig. 5.10). Indeed, the use of such expression elements significantly enhanced the vitamin E biosynthesis pathway in tobacco and tomato chloroplasts. In this study, there was a ten-fold increase in the accumulation of foreign proteins compared to the levels without IEEs.

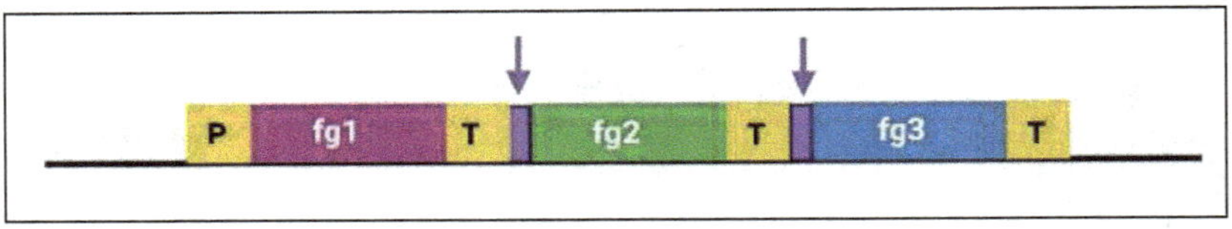

Fig. 5.10 A typical chloroplast vector comprises a promoter, P, foreign genes for various traits, fg1, fg2, fg3, and a transcription terminator, T. Arrows indicate the position of intercistronic expression elements, IEEs

Chloroplast transformation is typically achieved through particle bombardment. In general, gold particles are coated with the desired plasmid DNA, which after bombardment of the target tissue (e.g., leaf or embryonic stem), facilitates the delivery of the target genetic material into the chloroplast genome. Transgenic plants generated from the bombarded tissue are subjected to multiple rounds of selection using appropriate antibiotics. Due to the presence of potential multiple gene copies within a single cell, several regeneration cycles are required to obtain a stable transplastomic plantlet (Fig. 5.11).

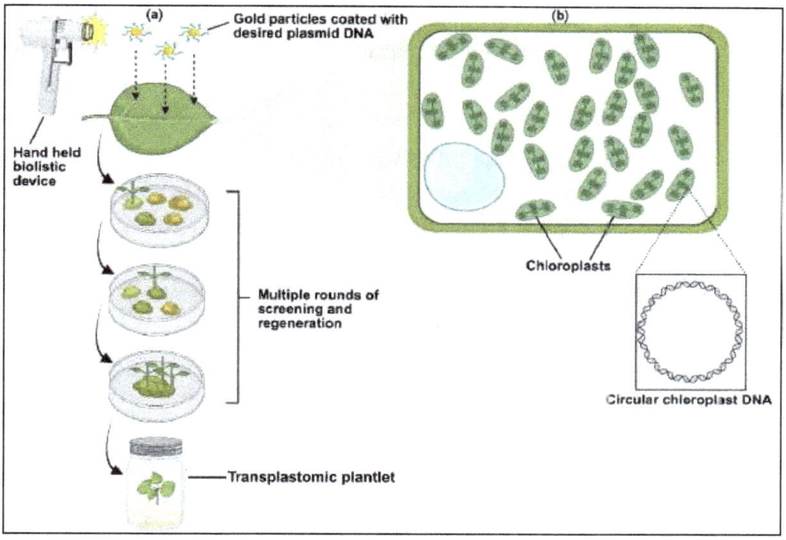

Fig. 5.11 (a) Schematic diagram showing the generation of a transplastomic plantlet, obtained after multiple rounds of screening and regeneration until the multi-copy chloroplast genome has been completely homogenized (b) plant cell showing multiple chloroplasts and double stranded chloroplast DNA

The *aad*A gene, encoding the enzyme 3'- adenylyl transferase, is the most widely used selectable marker gene for chloroplast transformation. It has strong resistance to the antibiotics spectinomycin and streptomycin, works on a variety of tissue types and provides an easy visual screen to identify transplastomic clones (Fig. 5.12).

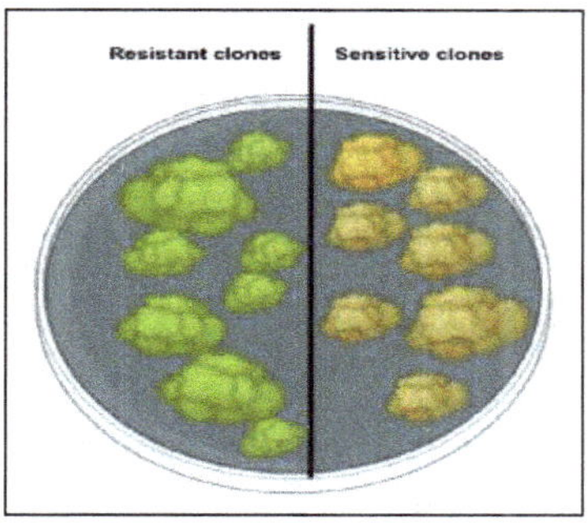

Fig. 5.12 Antibiotic resistance phenotype observed on an agar plate containing growth media with spectinomycin. Note that the antibiotic-resistant clones appear green while the sensitive clones appear chlorotic

When plant cells are grown on agar plates containing a selective antibiotic, chloroplasts carrying the transgenes are preferentially maintained. Over time, the transformed chloroplasts multiply until all copies of chloroplast DNA are uniformly transformed. Thus, it is possible to obtain a genetically stable transplastomic plant i.e., all chloroplast DNA copies are changed. Once a transplastomic plant is generated, the selective marker gene is no longer necessary and can be excised in many ways (see Section 5.7). Alternatively, the green fluorescent protein (*gfp)* (discussed in the next section) may be utilized as a selection marker. Transformed chloroplast expressing *gfp* will exhibit fluorescence when placed in the dark.

Since the development of chloroplast transformation in tobacco plants in the 1990s, there are currently many examples of chloroplast engineering in additional plant species (Table 5.2).

Table 5.2 Plant species in which chloroplast transformation is feasible

Family	Species	Common name
Solanaceae	*Nicotiana tabacum*	Tobacco
	Solanum lycopersicum	Tomato
	Solanum tuberosum	Potato
	Petunia hybrida var. Pink Wave	Petunia
	Scoparia melongena	Eggplant
	Capsicum annuum var. G4	Pepper
Brassicaceae	*Brassica oleracea*	Cauliflower
	Brassica capitate	Cabbage
	Brassica napus	Rapeseed
	Arabidopsis thaliana	Arabidopsis
	Lesquerella fendleri	Popweed
Fabaceae	*Glycine max*	Soybean
	Medicago sativa	Alfalfa
Salicaceae	*Populus alba*	Poplar
Amaranthaceae	*Beta vulgaris*	Sugar beet
Malvaceae	*Gossypium hirsutum*	Cotton
Poaceae	*Oryza sativa* var. Japonica line 19	Rice
Cucurbitaceae	*Momordica charantia*	Bitter squash

Achieving high expression levels of the desired (target) protein in the chloroplast may sometimes alter the plant's phenotype. For instance, a remarkably high expression level (making up to 40% of total soluble protein) of cellulolytic enzymes, which are useful for biofuel production, in tobacco chloroplast, resulted in retarded growth of transformed plants. Likewise, tobacco leaves expressing high levels of bacteria-phage lytic enzyme, plyGBS, were somewhat stunted compared to non-transformed tobacco plants. A variety of reasons have been cited for this: enzymatic activity of the recombinant protein impairs plastid-associated metabolic processes or negative interaction of the expressed protein with the chloroplast membranes or the high metabolic load imposed on the plants by the extreme accumulation levels of the recombinant protein. Future work is likely to focus on maximizing expression of proteins while maintaining a high level of plant growth.

Recent progress in chloroplast engineering includes efforts to enhance agronomic traits such as abiotic stress tolerance and pest resistance. One such approach is the use of RNAi technology for controlling insect pests in crops. Insect control through RNAi relies on plants that are manipulated to stably express double stranded RNA (dsRNA) or hair pin RNA (hpRNA) targeting essential insect genes. Chloroplasts are desirable targets to generate dsRNA because they lack the RNAi machinery. Therefore, when insects chew on leaves, they ingest the dsRNA, and these molecules are converted into small interfering RNAs or siRNAs which then bind to the complementary sequence of the insect's mRNA leading to the disruption in the expression of the target insect gene. Thus, essential physiological processes are impaired, resulting in growth retardation or death of the pest. It was observed that chloroplast RNA was taken up by insects (e.g., aphids or white flies) feeding on the phloem (plant sap). Thus, this system is effective in controlling phloem-feeding pests as well.

As expected, the higher accumulation of the target protein together with the desirable biocontainment, makes chloroplast engineering attractive over nuclear transformation. Further, development of selection markers, especially in plants that are naturally resistant to spectinomycin, and an efficient plant regeneration system are key to the further development of chloroplast expression systems. A serious limitation has been the poor expression of chloroplast genes in non-chloroplast tissues such as fruits, tubers and seeds. However, by using regulatory sequences derived from chloroplast genes with high expression in non-green tissues, it is possible to express chloroplast transgenes in any tissue type. A variety of regulatory elements have been tested and these may help with high transgene expression in non-green plastids.

5.4 Selectable marker genes and reporter genes

There are three aspects to creating transgenic plants. First, a process for introducing foreign DNA into plant cells, next a procedure for selecting those cells that have integrated the foreign DNA into the appropriate plant genome and finally a protocol for regeneration of plants from the transformed cell. A plant selectable marker provides a means of distinguishing transgenic material from non-transformed material. The selectable marker is usually co-transformed with the gene of interest. The selectable marker genes may be a toxic agent such as an antibiotic or herbicide, or they may be substrates for growth or that promote growth and cell differentiation of the transformed tissues. The selection marker gene codes for an enzyme with specificity for a particular substrate. An example of such a gene is NPTII (neomycin phosphotransferase II) gene, which confers resistance to the antibiotic kanamycin, and can result in the death of non-transformed plant cells in the presence of kanamycin. Kanamycin is the most frequently used antibiotic for selection of transformed plant cells; additionally in situations where NPTII was ineffective, hygromycin (linked to an HPT-hygromycin B phosphotransferase resistance gene) has often been used. Both NPTII and HPT are widely used selectable markers for nuclear transformation. On the other hand, the *aadA* gene, is a common selection marker for chloroplast transformation. It inactivates antibiotics like streptomycin and spectinomycin, as a result only plant cells that express the *aadA* gene can survive on medium containing these antibiotics, while non-transformed cells cannot. The marker genes that use antibiotics have been introduced into a variety of vectors and are broadly used in plant research and crop development; however, some of them are associated with biosafety concerns (see Section 5.7).

Reporter genes are useful to monitor transgenic events, and their activity can be assayed visually. Reporter genes assist in both the identification and quantification of gene expression by producing an identifiable and measurable response linked to gene activity. For an effective gene reporter system, it is vital that the reporter gene does not encode an enzyme activity that interferes with normal plant functions. Moreover, other factors such as enzyme stability, sensitivity, and specificity, as well as the quality, accuracy, and reproducibility of the assays are important determinants of the effectiveness of the reporter genes.

Scientists use reporter genes in two major contexts-

- *Transient expression*: This happens when the foreign gene does not integrate into the plant nuclear or chloroplast genome. Rather, it is expressed stably for a limited period in the plant cell cytoplasm. Scientists can quickly test whether the gene delivery method worked, saving time and resources.
- *Stable expression:* This happens when the gene integrates into the plant's genome, leading to inheritance and expression across generations.

Numerous genes have been tested as reporters for transformation, including genes that can be used as selectable markers and genes whose proteins produce a visible response to a specific assay (Table 5.3).

Table 5.3 Examples of selectable marker and reporter gene systems

Enzyme/protein encoded	Selectable marker	Reporter gene
Neomycin phosphotransferase II	Yes	No
Phosphinothricin *N*-acetyltransferase	Yes	No
Hygromycin B phosphotransferase	Yes	No
Dihydrofolate reductase	Yes	No
Aminoglycoside-3 adenyl transferase	Yes	No
5-Enolpyruvylshikimate-3-phosphate synthase	Yes	No
Chloramphenicol acetyltransferase	Yes	No
Gentamicin acetyltransferase	Yes	No
Threonine dehydratase	Yes	No
Blasticidin S deaminase	Yes	No
Nopaline synthase	No	Yes
Octopine synthase	No	Yes
β- Glucuronidase	No	Yes
β- Galactosidase	No	Yes
Firefly luciferase	No	Yes
Bacterial luciferase	No	Yes
Green fluorescent protein	No	Yes
Acetolactate synthase	Yes	No
Bromoxynil nitrilase	Yes	No
Xylose isomerase	Yes	No
Phosphomannose isomerase	Yes	No

Data from Glick, B.R. and Patten, C.L. 2022. "Molecular Biotechnology". 6th edition

Some common reporter genes that are used to monitor gene expression are discussed below.

β-glucuronidase (GUS): The *GUS* gene is one of the most popular reporter gene for transformation studies. In this system, the enzyme β-glucuronidase, hydrolyzes a wide range of β-glucuronides, many of which are commercially available as fluorometric or histochemical substrates. When plant tissues expressing the *GUS* gene are incubated with appropriate glucuronide-based substrates, the enzyme catalyzes the cleavage of the glucuronide bond, producing a readily detectable product — typically a colored, or fluorescent compound, which allows for easy visualization and quantification of gene expression. For example, GUS enzyme activity in plant extracts can be quantitatively measured by a fluorometric analysis that involves the hydrolysis of the substrate 4-methylumbelliferyl-β-D-glucuronide to produce fluorescent products.

A histochemical assay is particularly advantageous, in that GUS activity can be localized in individual cells of tissue sections. 5-bromo-4-chloro-3-indolyl β-D glucuronide (or X-Gluc) is a substrate used in histochemical GUS assays, where enzymatic cleavage by GUS produces an insoluble blue precipitate, indicating GUS activity in plant tissues (Fig. 5.13). However, this often requires sacrifice of the plant tissue making it unsuitable for selecting primary transformants or for long-term monitoring of gene expression. Nevertheless, the *GUS* gene is still widely used due to lack of intrinsic GUS activity in plants, the sensitivity with which it can be measured, and it is not easily degraded in plant tissues.

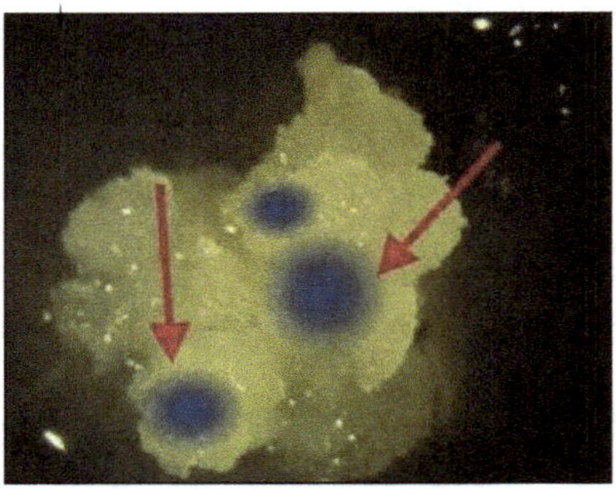

Fig. 5.13 Transformed callus tissue incubated with X-Gluc substrate to produce blue staining at the site of GUS enzyme activity (red arrows)

Green fluorescent protein (GFP): The green fluorescent protein (GFP) (isolated from the jellyfish *Aequorea victoria*) is a versatile marker for *in vivo* studies. The process involves the incorporation of the *gfp* gene into the DNA construct or the vector, which allows transformed material to fluoresce bright green under UV light. It provides the potential for screening transformants at the early stage of selection without the need for tissue destruction or chemical staining, unlike GUS. Moreover, it does not require any substrates or co-factors. Furthermore, it can be fused with other proteins of interest without interfering with the plants' basic functions. A highly sensitive imaging equipment, like a fluorescence microscope (Fig. 5.14a), can be used for real time imaging of live cells and a wide variety of transformed material i.e., embryos, cotyledons, leaves, roots, leaf discs, meristems, callus tissue, and seeds. A hand-held UV lamp (Fig. 5.14b), is a convenient and cost-effective alternative to a microscope, to monitor transformed plants grown in tissue culture, in the greenhouse or in the field.

Currently, engineered variants of GFP for improved plant expression or altered spectral properties have significantly improved its utility as a reporter gene.

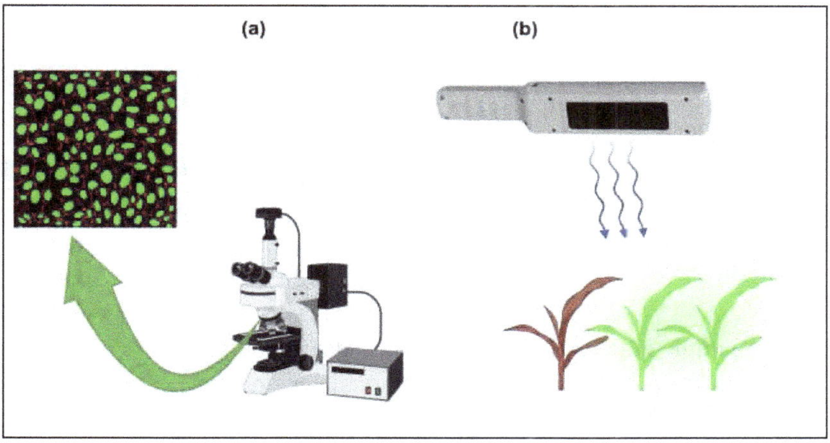

Fig. 5.14 (a) Fluorescence microscope, (b) hand-held UV lamp; *gfp*-expressing plants show green fluorescence, while non-transformed leaves show red autofluorescence

Luciferase: The two most common luciferases which have been widely used as reporter genes are the firefly (*Photinus pyralis*) and bacterial (*Vibrio fischeri*) luciferases. In presence of the luminescent substrate luciferin, it emits light. The luciferase signal can be measured by a luminometer. Luciferase offers the advantage of monitoring gene expression non-destructively, but it does require a special device and a substrate. Moreover, luciferase is a relatively unstable gene, so the signal can be short lived. Although, it is a good option for real-time, rapid, and highly sensitive and non-destructive detection of gene expression, it may not be effective for long-term studies because its signal fades quickly once the enzyme degrades. Nonetheless, luciferase continues to be a popular reporter gene due to its non-invasive nature.

5.5 Targeted genome editing

The most widely used system for transformation of plants is via the Ti-plasmid of *A. tumefaciens*. However, as previously stated, this method has some disadvantages in that the transgene integration occurs at random locations in the host genome. As a result, only a minor fraction of the transformed plants will actually express the desired trait. This makes the screening and selection process cumbersome and time consuming. In contrast, targeted DNA insertion at specifically selected genomic sites increases the chances of creating the desired traits in the resulting plants.

In the 1990s, DNA-cutting enzymes such as the meganucleases (also called homing endonucleases) were used to insert *Agrobacterium*-transformed DNA at specific sites in the plant genome. Meganucleases recognize and cleave DNA at genomic targets with specific sequences (14bp-40bp) and create a double stranded break. One problem with their use is that they can only target sites that already exist in the genome. The availability of suitable targets is therefore limited. In contrast, newer editing tools allow greater flexibility. For instance, zinc finger nucleases (ZFNs), are artificially designed proteins consisting of a DNA-binding domain (multiple zinc finger repeats), and a cleavage domain which is derived from the bacterial enzyme *Fok*I. The zinc finger domains are engineered to recognize chosen DNA sequences of 9–18 base pairs, enabling modification at virtually any genomic location. The *Fok*I nuclease domain can only cut DNA when the enzyme is dimerized, requiring two ZFNs to bind to opposite strands of the DNA flanking the target site. While they are still somewhat useful, ZFNs have now been superseded by transcription activator-like effector nucleases (TALENs) and CRISPR/Cas9 systems and the range of applications for these techniques appear to be relatively unlimited. Like ZFNs, TALENs consist of customizable TALE DNA binding domains fused with non-specific *Fok*I DNA-cutting domains (Fig. 5.15). TALEs consist of a series of 34-amino-acid repeats, each of which recognizes a specific stretch of DNA sequence. Within each repeat, two variable amino acids determine the DNA base specificity. Thus, with this approach one can customize the order of repeats to bind any DNA sequence, base by base.

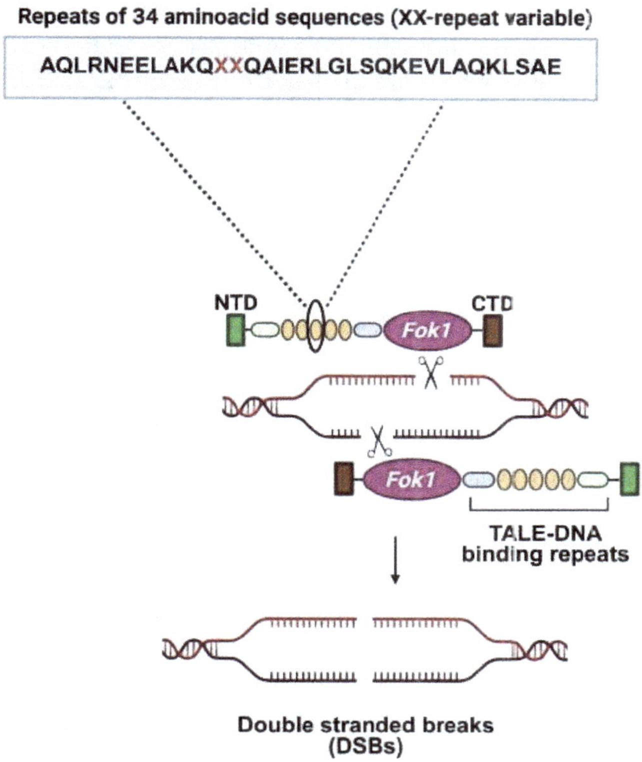

Fig. 5.15 TALEN, the scissor enzyme. NTD and CTD are N-and C-terminal domains respectively, flanking the DNA-binding TALE domain. The spacer sequence separating the two TALE-binding sequences, TALEN induced double stranded breaks (DSBs) are also shown

As previously mentioned, targeted gene insertion by CRISPR/Cas9 is another approach for precisely engineering the plant genome. The CRISPR/Cas9 system has advanced rapidly since it was first used for gene targeting in plants in 2013. Since then, it has seen explosive growth, with numerous studies highlighting its simplicity and efficiency. It is simple and low in cost compared to ZFNs or TALENs, as it does not require protein engineering and development. The CRISPR/Cas9 system relies on DNA-RNA interaction for target DNA sequence recognition, and therefore is a relatively facile genome editing tool. One potential problem of this approach is that Cas9 sometimes induces DSBs at unintended locations in the genome, leading to an unpredictable outcome. However, this problem has been addressed by engineering high-fidelity Cas9 proteins (e.g., SpCas9 protein) which binds more accurately to the target DNA sequence. One group of researchers showed that the engineered SpCas9 can recognize a broader range of PAM (protospacer adjacent motifs, a short stretch of DNA that is required for Cas9 to bind targets and cut) sequences allowing the flexibility to edit more locations in the genome.

The capacity to target multiple genes simultaneously for editing has facilitated the development of many elite plant cultivars. For instance, in wild cherry tomato, scientists used CRISPR/Cas9 to edit three genes that control plant architecture, flowering time and growth duration. The genome edited tomato plants demonstrated condensed shoots, rapid flowering and a shorter life cycle, which are all valuable traits in growing crops in an urban environment. Plants with compact growth are ideal for vertical set ups because they take up less space.

Currently, CRISPR/Cas9-based technologies use *Agrobacterium*-mediated delivery or microprojectile bombardment of callus cultures. Subsequently, robust regeneration protocols are needed to produce genome edited plants. However, tissue culture can only be applied to a limited number of plant species and is generally both labor intensive and time consuming. A technique that avoids the use of conventional tissue culture procedures involves the *de novo* induction of meristems (Fig. 5.16). This can be achieved through the delivery of an *Agrobacterium* strain harboring a vector expressing the CRISPR/Cas9 editing system and regulatory genes involved in plant morphogenesis. This strategy leverages the totipotency of plant cells leading to plant regeneration of gene-edited individuals directly from somatic tissues. This method appears to hold promise in editing plant cultivars that are recalcitrant to tissue culture, potentially improving transformation efficiency and broadening the range of readily manipulable plant species.

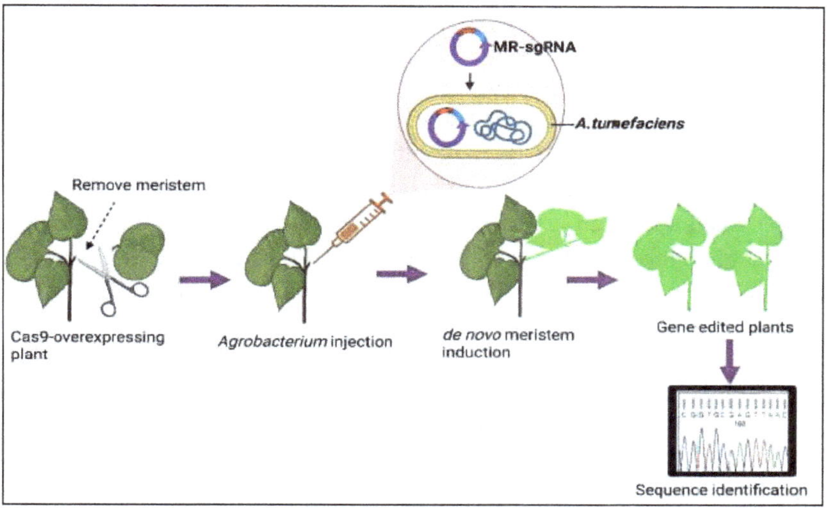

Fig. 5.16 Scheme showing genome editing by *de novo* induction of a meristem. Removal of an existing meristem in a plant expressing the Cas9 protein, the cut site is injected with an *Agrobacterium tumefaciens* culture expressing MR (morphogenic regulator) and the guide RNA. Over time, gene edited meristems form, resulting in edited plants. Subsequently, plants can be confirmed for successful gene editing by DNA sequencing of the altered region of the plant genome

For editing, organellar genomes, such as those of a chloroplast, it is required that both the Cas9 protein and the guide RNA (gRNA) are transferred and expressed within the organelle itself. Recently, a group of scientists reported that the efficiency of chloroplast transformation in tobacco plants increased up to ten-fold by using the CRISPR/Cas9 system. The gRNA and Cas9 were added to the chloroplast transformation vector that includes homologous flanking regions (Fig. 5.17). This procedure enabled the precise integration of exogenous DNA and targeted cleavage of the chloroplast genome through homologous recombination.

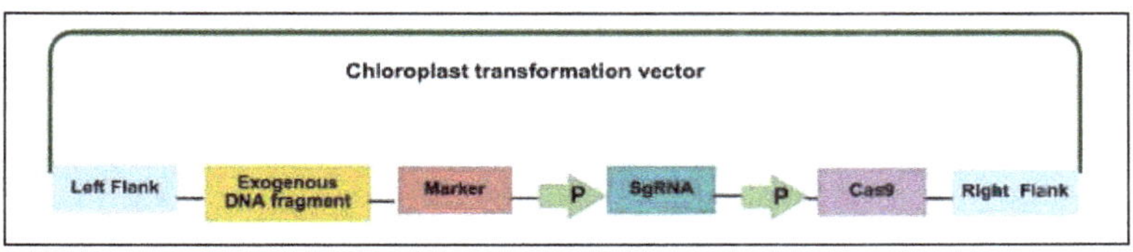

Fig. 5.17 Schematic representation of a CRISPR/Cas9 construct for editing the chloroplast genome. Promoter, P, marker, exogenous DNA, left and right flanking sequences are also shown

5.6 Manipulating protein glycosylation

Glycosylation is a post-translational modification where carbohydrates (glycans) are enzymatically attached to proteins. The functions and stability of many therapeutic proteins (hormones, antibodies or enzymes) can be affected by the addition of glycans to specific amino acid residues. Importantly, glycosylation in mammalian and plant cells is covalently N-linked via an asparagine amide. However, the type of outer core carbohydrate chains attached to glycosylated proteins in plants are different from those found in mammalian proteins. Although, these differences may not directly alter the activity of a protein, if a protein is produced for therapeutic applications, incorrect glycosylation may cause immunogenicity in humans or alter antibody effector functions, thus limiting the use of the protein. Consequently, modification of N-linked glycans to more closely resemble the modifications found in mammalian cells can significantly improve the quality features of final products

Two independent studies used RNA interference strategy to produce human monoclonal antibodies (Mabs) and human like N-glycans in plants. In these studies, the researchers generated transgenic tobacco and duckweed (*Lemna minor*) plants by knocking out α-1,3-fucosyltransferase (α-1,3-FucT) and β-1,2-xylosyltransferase (XylT) genes. In both studies, the resultant Mab contained a single glycan core structure without any plant-specific N-glycan residues (Fig. 5.18). Moreover, these Mabs exhibited superior qualities i.e., increased antibody dependent cellular cytotoxicity activity and antigen-binding activity compared to the Mab synthesized in Chinese hamster ovary (CHO) cells in culture. Indeed, plants can serve as promising platforms for the production of human therapeutics, and can significantly lower production costs compared to CHO cells, especially when using aquatic plants such as duckweed, which is a non-food and non-feed crop and can be cultivated without the need for arable land.

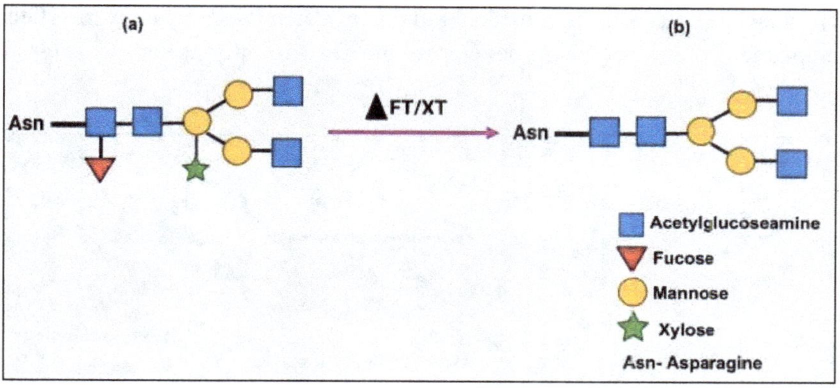

Fig. 5.18 Structures of (a) native and (b) plant-engineered glycoforms

While plants typically add plant-specific β-1,2-xylose and α-1,3-fucose to N-glycans, in carrot cells, this can be avoided by taking advantage of their natural tendency to produce glycans rich in mannose. This aspect was leveraged in the production in carrot cells of recombinant glucocerebrosidase (GCD), as a replacement therapy for the treatment of Gaucher disease (a rare lysosomal storage disorder). The carrot-derived GCD is targeted to the storage vacuoles, using a plant-specific C-terminal signal sequence (Fig. 5.19).

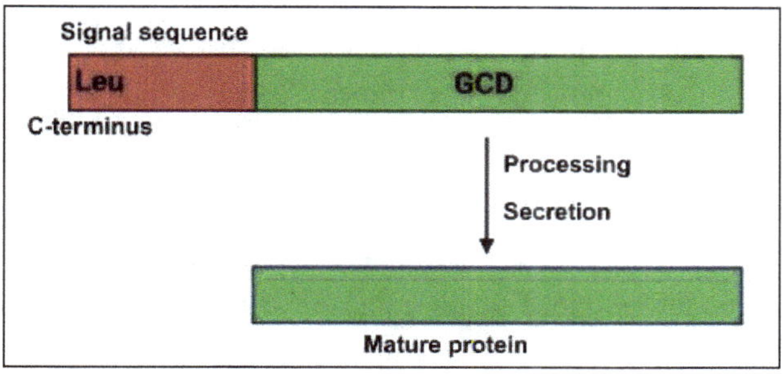

Fig. 5.19 Schematic illustration showing the secretion of GCD using a C-terminal vacuolar targeting signal sequence

As a result, the trimming of glycans exposes the terminal mannose residues, and yields a protein with similar efficacy compared to the enzyme produced in CHO cells. Whereas the recombinant GCD produced in mammalian cells needs an enzymatic trimming to expose the mannose residues (Fig. 5.20), potentially increasing the cost of production. It is important to note that in clinical trials the plant-specific glycans were not problematic in the carrot-made GCD, however their presence still requires consideration due to potential safety concerns. Other therapeutic proteins are being developed and are currently at various stages of clinical trials.

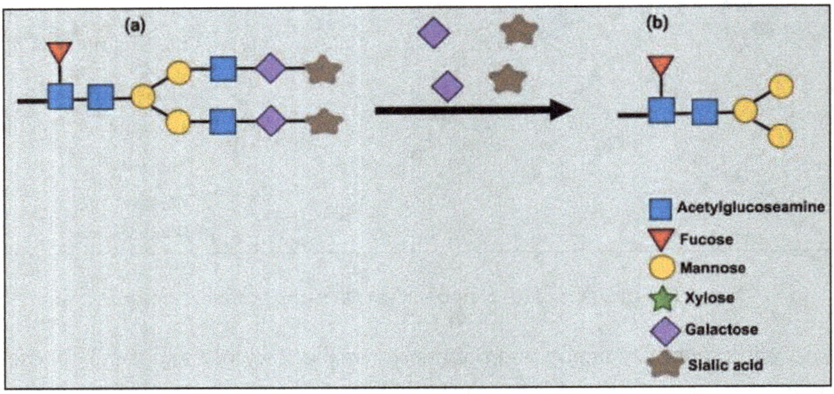

Fig. 5.20 Major glycan structure of GCD expressed in Chinese hamster ovary cells (a) before (b) after enzymatic deglycosylation

5.7 Creating marker-free transgenic plants

The main function of the selectable marker is to assist in the identification of transformants. Commonly, the selectable marker gene is no longer needed once the transgenic plant is generated, but the presence of the gene in the final plant has caused regulatory and public concern due to the following reasons:

i. The products of marker genes may be toxic, allergenic or pathogenic.
ii. Antibiotic resistance genes could be transferred into microorganisms and increase the number of resistant pathogenic microorganisms in the environment or the human gut.
iii. Gene transfer to other organisms via consumption of plant products is possible, although highly unlikely.
iv. A transgenic plant that already contains a specific selectable marker gene cannot be transformed again using the same marker.
v. Resistance genes may escape to wild relatives.

Several strategies have been developed to eliminate antibiotic-resistant markers in the final products, and some of them are discussed below.

Removal of marker genes from nuclear DNA

Co-transformation method

In this method the target gene and the marker gene are introduced into the plant at the same time, using two separate vectors. Generally, the co-transformation frequencies are high (~80%) with this approach and the transformed plants will contain both genes, often integrated at different sites on the chromosomal DNA. Through repeated genetic crosses, the marker genes can be segregated away from the target gene. Thus, transgenic marker free plants can be obtained this way (Fig. 5.21). This approach, while effective, involves laborious tissue culture and regeneration steps, often not useful for sterile plants (no sexual reproduction), vegetatively propagated species (asexual reproduction), or woody plants (long generation times).

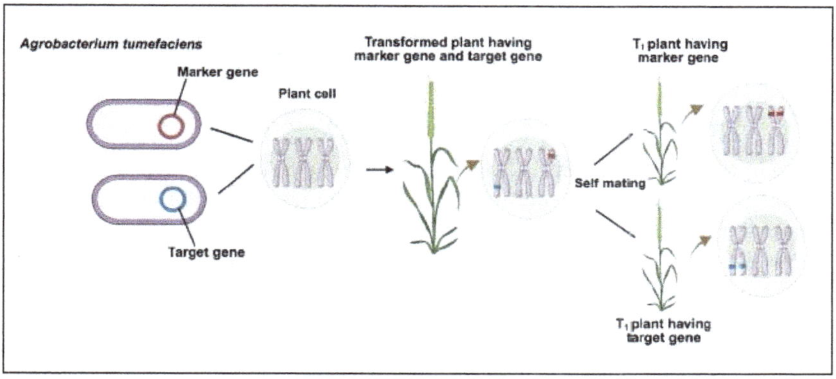

Fig. 5.21 The selective marker gene and the target gene can be introduced either on separate vectors on a single *Agrobacterium* strain, or on the same vector. Alternatively, the two genes can be delivered by particle bombardment

Transposon-based method

In this method, the selectable marker gene is placed on a transposable element (Ds) and then later excised from the genome after the expression of the enzyme transposase. In fact, any sequence that is between two Ds elements will be mobilized to a new location in the genome. For this approach, two different strategies may be used. In the first instance, the target gene is inserted between the Ds repeat elements (Fig. 5.22a). In this scenario, following the

Agrobacterium-catalyzed transformation, the target gene will be excised and typically reinserted at a genetic locus far away from the locus in which the selectable marker gene is located, so that they can subsequently be segregated following self-mating of the transformants. In the second approach the marker gene is flanked by the Ds repeat elements (Fig. 5.22b) with the marker gene being segregated following self-mating of the transformants. Both approaches have been applied but since re-insertion frequencies are high, the second approach is not very efficient.

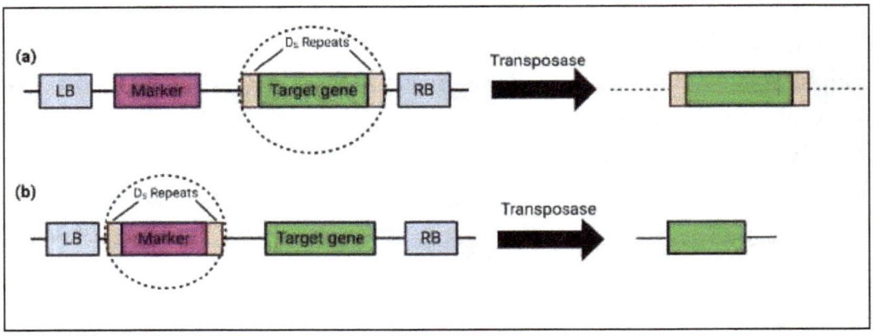

Fig. 5.22 Transposon based methods for marker gene elimination from the nuclear genome of transgenic plants. The activation of transposase leads to (a) excision of the target gene, its reinsertion at a different genomic location and its subsequent segregation from the marker gene in the progeny (b) the marker gene 'jumps out' onto a different genomic location from the target gene or is lost

Transposon-based methods help remove marker genes and create plants with transgenes at different positions from the original transformant. This is especially useful for recalcitrant plants, as having a set of plants with the same gene but at different loci increases the chances of finding a line with good, stable expression. However, considering that this procedure also uses sexual crossing, the segregation of transgenes is not feasible in vegetatively propagated crops.

To circumvent problems related to development of marker free plants without sexual crossing is to flank the selectable marker with specific DNA sequences. This may be done by utilizing site-specific recombinases which have the ability to cleave DNA at specific sites. Many site-specific recombinases of yeast or prokaryotic origin have been used for marker removal in transgenic plants. The cognate recombinase enzyme (Cre) from *E. coli* bacteriophage P1, recognizes a 34-bp site called *lox*P. Cre-mediated recombination between two *lox*P sites, leads to the excision of the intervening selectable marker sequences

(Fig. 5.23). Marker-free plant lines are produced by the *in vitro* regeneration of modified cells under selection, to confirm the presence and expression of the target gene, and then grown in culture until complete excision of the selectable marker is achieved. Because of a relatively short sequence length i.e., 34 bp, it can sometimes result in unexpected excision of the marker gene due to an inactive Cre. However, the application of such mechanisms are unlimited and numerous excision systems have been developed.

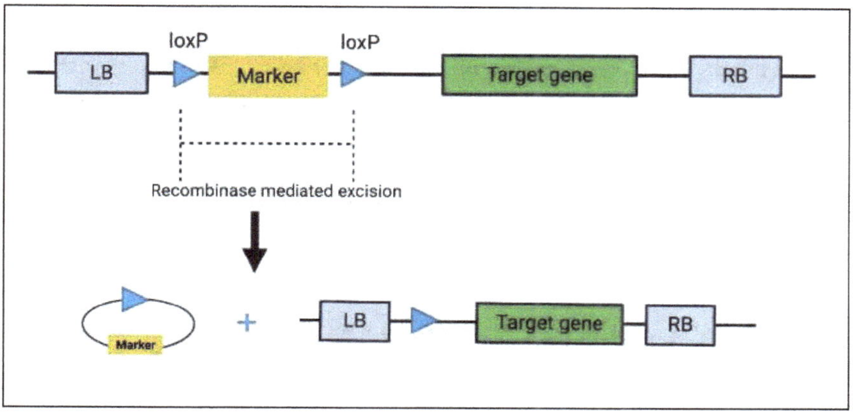

Fig. 5.23 Illustration of recombinase-mediated selectable marker removal. The marker gene is flanked by loxP sequences, recognized by Cre recombinase. The excision event removes the marker between the associated recognition sites leaving the target gene intact. The non-replicating circular DNA fragment is lost

Removal of marker genes from chloroplast DNA

The high expression of transgenes is achievable in chloroplasts due to the high copy number of the chloroplast genomes. This factor generally makes chloroplast transformation a more preferred technique for introducing a foreign gene with an intended high level of protein expression. However, when antibiotic resistance genes used as selection markers are co-introduced along with the foreign gene, their expression is also amplified. For example, thousands of copies of the *aad*A gene, which confers resistance to the antibiotics spectinomycin and streptomycin, are present per cell in transplastomic plants. This obviously increases the potential risk of *aad*A gene transfer from plants to the environment, which makes its removal necessary.

Marker genes may be removed through the homologous recombination mechanism in the chloroplast genome. Generally, the marker gene is flanked by directly repeating sequences (174 bp) and introduced along with the foreign gene into the genomic sequence of the vector. Following introduction into the chloroplast, the chloroplast DNA sequences direct the integration of the foreign gene into the chloroplast genome by homologous recombination with the flanking left and right regions. The removal of antibiotic selection causes the marker gene to be subsequently lost during a second recombination event. Thus, after the initial transformants are selected by spectinomycin resistance, the transplastomic plantlets are grown without spectinomycin, and the selectable marker gene is lost as a result of the homologous recombination via the two identical repeat sequences (Fig. 5.24).

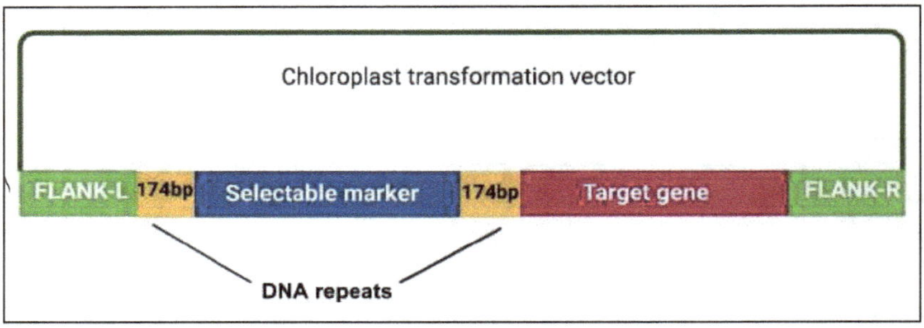

Fig.5.24 Schematic representation of a selectable marker gene removal system for a transformed chloroplast. During chloroplast transformation, the selectable marker gene and the target gene are incorporated in the chloroplast genome by homologous recombination with chloroplast DNA flanking the entire integrating DNA. Transformants are initially selected based on their resistance to spectinomycin. Then, transformants are grown in the absence of antibiotics, and the selectable antibiotic resistance marker gene is lost due to homologous recombination between the 174-bp repeat elements

Alternatively, use of the betaine aldehyde dehydrogenase gene, naturally present in spinach and a few other plant species, as a selectable marker has facilitated development of transgenic plants without the use of antibiotic selection. This enzyme converts toxic betaine aldehyde into non-toxic glycine betaine. Untransformed plants are killed when grown in a medium containing betaine aldehyde, thus allowing selection of only transformed chloroplasts to survive on medium containing betaine aldehyde. Additionally, glycine betaine confers abiotic stress tolerance upon the transformed plant, offering dual benefits. This strategy enabled successful regeneration of tomato and tobacco plants, and might lead to the development of transgenic crops, in particular those that are naturally resistant to spectinomycin. Other studies have used less toxic selection agents, with a specificity for a different enzyme. An example of such an agent is the mannose utilizing gene, mannose phosphate isomerase (MPI). When mannose was used as a substrate plants transformed with MPI were able to grow, whereas non-transformed tissues stop growing because they cannot use mannose for energy. Positive selection based on growth of transformed tissue has also been reported when the GUS gene was adapted to function as a selectable marker gene. In this case, selection of transformants was achieved by the use of the substrate benzyladenine-glucuronide which is hydrolyzed by GUS to produce benzyladenine, a cytokinin. Cytokinins are plant growth hormones that stimulate shoot formation and plants that have been engineered with such a selectable marker are likely to show enhanced shoot regeneration. In fact, it was demonstrated in tobacco that the regeneration capacity of transformed cells using benzyladenine-glucuronide was 3-fold higher than when transformants were selected with kanamycin. A GFP protein has been modified for use as a selection marker. The modified version called eYGFPuv, was effectively used to differentiate transgenic *Arabidopsis* seedlings, which were easily identified by the green fluorescence they exhibited under UV light. Although currently limited to transforming floral tissue, the successful transformation without the requirement for antibiotics represents an encouraging result.

For commercialization of transgenic crops, the antibiotic marker genes must be removed to address the aforementioned concerns. A number of methods now exist to achieve this.

5.8 Further readings

Belaffif, Mohammad B, Morgan C Brown, Brenda Marcial, Can Baysal, and Kankshita Swaminathan. 2025. "New Strategies to Advance Plant Transformation." *Current Opinion in Biotechnology* 91 (February):103241. https://doi.org/10.1016/j.copbio.2024.103241.

Bock, Ralph. 2014. "Engineering Chloroplasts for High-Level Foreign Protein Expression." In *Chloroplast Biotechnology*, edited by Pal Maliga, 1132:93–106. Methods in Molecular Biology. Totowa, NJ: Humana Press. https://doi.org/10.1007/978-1-62703-995-6_5.

Bortesi, Luisa, and Rainer Fischer. 2015. “The CRISPR/Cas9 System for Plant Genome Editing and Beyond.” *Biotechnology Advances* 33 (1): 41–52. https://doi.org/10.1016/j.biotechadv.2014.12.006.

Chen, Zongliang, Juan M. Debernardi, Jorge Dubcovsky, and Andrea Gallavotti. 2022. “Recent Advances in Crop Transformation Technologies.” *Nature Plants* 8 (12): 1343–51. https://doi.org/10.1038/s41477-022-01295-8.

Chiu, Wan-ling, Yasuo Niwa, Weike Zeng, Takanori Hirano, Hirokazu Kobayashi, and Jen Sheen. 1996. “Engineered GFP as a Vital Reporter in Plants.” *Current Biology* 6 (3): 325–30. https://doi.org/10.1016/S0960-9822(02)00483-9.

Cox, Kevin M, Jason D Sterling, Jeffrey T Regan, John R Gasdaska, Karen K Frantz, Charles G Peele, Amelia Black, et al. 2006. “Glycan Optimization of a Human Monoclonal Antibody in the Aquatic Plant Lemna Minor.” *Nature Biotechnology* 24 (12): 1591–97. https://doi.org/10.1038/nbt1260.

Dale, E C, and D W Ow. 1991. “Gene Transfer with Subsequent Removal of the Selection Gene from the Host Genome.” *Proceedings of the National Academy of Sciences* 88 (23): 10558–62. https://doi.org/10.1073/pnas.88.23.10558.

Dale, Philip J. 1992. “Spread of Engineered Genes to Wild Relatives.” *Plant Physiology* 100 (1): 13–15. https://doi.org/10.1104/pp.100.1.13.

Daniell, H, J Vivekananda, B L Nielsen, G N Ye, K K Tewari, and J C Sanford. 1990. “Transient Foreign Gene Expression in Chloroplasts of Cultured Tobacco Cells after Biolistic Delivery of Chloroplast Vectors.” *Proceedings of the National Academy of Sciences* 87 (1): 88–92. https://doi.org/10.1073/pnas.87.1.88.

Darbani, Behrooz, Amin Eimanifar, C. Neal Stewart, and William N. Camargo. 2007. “Methods to Produce Marker-free Transgenic Plants.” *Biotechnology Journal* 2 (1): 83–90. https://doi.org/10.1002/biot.200600182.

Dong, Oliver Xiaoou, and Pamela C. Ronald. 2021. “Targeted DNA Insertion in Plants.” *Proceedings of the National Academy of Sciences* 118 (22): e2004834117. https://doi.org/10.1073/pnas.2004834117.

El-Mounadi, Kaoutar, María Luisa Morales-Floriano, and Hernan Garcia-Ruiz. 2020. “Principles, Applications, and Biosafety of Plant Genome Editing

Using CRISPR-Cas9." *Frontiers in Plant Science* 11 (February):56. https://doi.org/10.3389/fpls.2020.00056.

Gelvin, Stanton B. 2012. "Traversing the Cell: Agrobacterium T-DNA's Journey to the Host Genome." *Frontiers in Plant Science* 3. https://doi.org/10.3389/fpls.2012.00052.

Glick, Bernard R. 1995. Metabolic load and heterologous gene expression. *Biotechnology Advances* 13: 247-261.

Glick, Bernard R., and Cheryl L. Patten. 2022. *Molecular Biotechnology: Principles and Applications of Recombinant DNA*. Sixth edition. Washington, DC: ASM Press.

Goldsbrough, Andrew. 1992. "Marker Gene Removal: A Practical Necessity?" *Trends in Biotechnology* 10:417. https://doi.org/10.1016/0167-7799(92)90289-8.

He, Yubing, Tao Zhang, Hui Sun, Huadong Zhan, and Yunde Zhao. 2020. "A Reporter for Noninvasively Monitoring Gene Expression and Plant Transformation." *Horticulture Research* 7 (1): 152. https://doi.org/10.1038/s41438-020-00390-1.

Heifetz, P. 2001. "Protein Expression in Plastids." *Current Opinion in Plant Biology* 4 (2): 157–61. https://doi.org/10.1016/S1369-5266(00)00153-9.

Hopp, Horacio Esteban, German Spangenberg, and Luis Herrera-Estrella. 2022. "Editorial: Plant Transformation." *Frontiers in Plant Science* 13 (April):876671. https://doi.org/10.3389/fpls.2022.876671.

Iamtham, Siriluck, and Anil Day. 2000. "Removal of Antibiotic Resistance Genes from Transgenic Tobacco Plastids." *Nature Biotechnology* 18 (11): 1172–76. https://doi.org/10.1038/81161.

Lu, Yinghong, Habib Rijzaani, Daniel Karcher, Stephanie Ruf, and Ralph Bock. 2013. "Efficient Metabolic Pathway Engineering in Transgenic Tobacco and Tomato Plastids with Synthetic Multigene Operons." *Proceedings of the National Academy of Sciences* 110 (8). https://doi.org/10.1073/pnas.1216898110.

Mahfouz, Magdy M., Agnieszka Piatek, and Charles Neal Stewart. 2014. "Genome Engineering via TALENs and CRISPR/Cas9 Systems: Challenges and

Perspectives." *Plant Biotechnology Journal* 12 (8): 1006–14. https://doi.org/10.1111/pbi.12256.

Mastrangeli, Renato, Wolf Palinsky, and Horst Bierau. 2019. "Glycoengineered Antibodies: Towards the next-Generation of Immunotherapeutics." *Glycobiology* 29 (3): 199–210. https://doi.org/10.1093/glycob/cwy092.

Mao, Yanfei, Jose Ramon Botella, Yaoguang Liu, and Jian-Kang Zhu. 2019. "Gene Editing in Plants: Progress and Challenges." *National Science Review* 6 (3): 421–37. https://doi.org/10.1093/nsr/nwz005.

Maher, Michael F., Ryan A. Nasti, Macy Vollbrecht, Colby G. Starker, Matthew D. Clark, and Daniel F. Voytas. 2020. "Plant Gene Editing through de Novo Induction of Meristems." *Nature Biotechnology* 38 (1): 84–89. https://doi.org/10.1038/s41587-019-0337-2.

Miki, Brian and McHugh, Sylvia. 2004. "Selectable Marker Genes in Transgenic Plants: Applications, Alternatives and Biosafety" *Journal of Biotechnology* 107 (3):193-232. https://doi.org/10.1016/j.jbiotec.2003.10.011.

Narra, Muralikrishna, Issei Nakazato, Brittany Polley, Shin-ichi Arimura, Grant N. Woronuk, and Pankaj K. Bhowmik. 2025. "Recent Trends and Advances in Chloroplast Engineering and Transformation Methods." *Frontiers in Plant Science* 16 (April):1526578. https://doi.org/10.3389/fpls.2025.1526578.

Ozyigit, Ibrahim Ilker, and Kuaybe Yucebilgili Kurtoglu. 2020. "Particle Bombardment Technology and Its Applications in Plants." *Molecular Biology Reports* 47 (12): 9831–47. https://doi.org/10.1007/s11033-020-06001-5.

Shaaltiel, Yoseph, Daniel Bartfeld, Sharon Hashmueli, Gideon Baum, Einat Brill-Almon, Gad Galili, Orly Dym, et al. 2007. "Production of Glucocerebrosidase with Terminal Mannose Glycans for Enzyme Replacement Therapy of Gaucher's Disease Using a Plant Cell System." *Plant Biotechnology Journal* 5 (5): 579–90. https://doi.org/10.1111/j.1467-7652.2007.00263.x.

Stegelmeier, Ashley A., Danielle M. Rose, Benjamin R. Joris, and Bernard R. Glick. 2022. "The Use of PGPB to Promote Plant Hydroponic Growth." *Plants* 11 (20): 2783. https://doi.org/10.3390/plants11202783.

Sugita, K., E. Matsunaga, and H. Ebinuma. 1999. "Effective Selection System for Generating Marker-Free Transgenic Plants Independent of Sexual Crossing." *Plant Cell Reports* 18 (11): 941–47. https://doi.org/10.1007/s002990050688.

Vain, P., J. De Buyser, V. Bui Trang, R. Haicour, and Y. Henry. 1995. "Foreign Gene Delivery into Monocotyledonous Species." *Biotechnology Advances* 13 (4): 653–71. https://doi.org/10.1016/0734-9750(95)02009-8.

Wang, Yueju, Yuan-Yeu Yau, Donna Perkins-Balding, and James G. Thomson. 2011. "Recombinase Technology: Applications and Possibilities." *Plant Cell Reports* 30 (3): 267–85. https://doi.org/10.1007/s00299-010-0938-1.

Chapter 6

Useful Transgenic Plants

Since scientists first began to genetically engineer plants around 40 or so years ago, a number of transgenic plants have been commercialized and many more transgenic plants are in various stages of development. Commercialized transgenic plants include those that are insect resistant, virus resistant, and herbicide resistant. Transgenic plants that have been engineered but not yet commercialized include those that: are resistant to fungal and bacterial pathogens; help plants to overcome salt and drought stress; slow fruit ripening and flower wilting; produce medicinal compounds such as pharmaceuticals; synthesize antibodies, and vaccines; modify food taste and/or appearance; increase plant yield; and increase plant fruit, vegetable and/or seed nutritional content. With the many hundreds of transgenic plants being produced every year, more and more of these are being tested, approved and then commercialized.

6.1 Insect resistance

Bacillus thuringiensis (*Bt*)

Most economically important crops suffer significant yield losses due to noxious insects. A variety of natural defense mechanisms exist in plants to deter insect pests (see Chapter 2); yet these mechanisms are not always sufficient to prevent pest damage at acceptable levels for crop plants. Therefore, one solution to this problem is to use chemical insecticides. However, increasing concern about contamination of the environment, toxicity to animals and the persistence of residues from chemical insecticides in soil and water have encouraged the development of alternative approaches of controlling insect pests. The spore-forming Gram-positive bacterium *Bacillus thuringiensis* (*Bt*) produces a diverse range of insecticidal proteins (*Bt* toxins), which are found to be toxic to different groups of insect pests. Some advantages of *Bt* toxins compared to chemical insecticides include an exceptional level of specificity towards targeted pests and extremely low mammalian toxicity. Additionally, *Bt* toxins are biodegradable and have not been associated with any detrimental effects on non-target species such as humans, farm animals and birds. Many products that include *Bt* toxins have been on the market since 1980. In particular, *Bt* subspecies *Kurstaki* and *Aizawa*, have received considerable attention and are applied as foliar sprays, much like synthetic insecticides, for managing insect-infested plantations in agriculture and forestry. In many instances *Bt* toxins significantly have reduced or replaced the use of chemical insecticides.

The insecticidal activities of *Bt* toxins reside in the crystal (Cry) proteins, which are synthesized during the bacterial sporulation (stationary) phase of its growth cycle. In fact, the Cry proteins are part of a larger structure called a parasporal crystal, which upon ingestion by the insect pest, are converted from a protoxin to an active toxin form in the insect midgut. This process happens in the presence of specific insect digestive proteases. In its active form, the toxic proteins then bind to specific receptors of the gut epithelial cells of the target insect and form pores in the cell membrane, thwarting cell metabolism, and resulting in cell lysis. Within a few hours of ingesting the Cry protein, the insect becomes dehydrated, and eventually dies (Fig. 6.1). Perhaps, the most important aspect of this technology is the selective toxicity of *Bt* protoxins, since they enable efficient control of target pests without the undesired effects on non-target organisms. Using recombinant DNA technology, scientists have developed new *Bt* strains for more effective pest control in different crops.

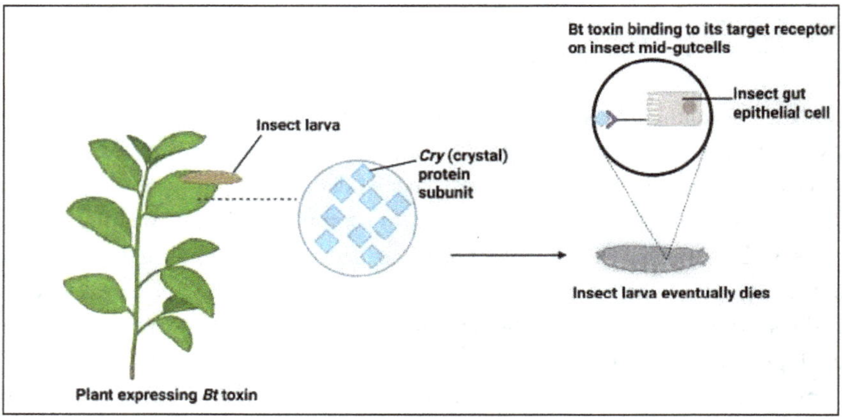

Fig. 6.1 An insect larva feeding on a plant leaf expressing the *Bacillus thuringiensis* (*Bt*) parasporal crystal (insecticidal protoxin). The insect larva dies 2-5 days after protoxin ingestion. The activated protein binds specifically to receptors in the mid gut of the insect and brings about lysis of these cells by formation of pores, eventually leading to the death of the insect larva

In addition to its use as an insecticidal spray, *Bt* has been successfully utilized as a source of *Cry* genes to genetically engineer insect resistance into crop plants. Plants modified in this way may be obtained by using common gene transfer techniques. Large scale use of insect-resistance crops began in 1995, e.g., following regulatory approval, *Bt* cotton was planted in 1995, followed by *Bt* corn. Genes for several of these insecticidal proteins, active against several different insects have been expressed in a wide variety of crops (Table 6.1).

Table 6.1 List of several commercialized transgenic plants expressing a wide range of Cry-protein genes from *Bt*

Transformed plant	Insecticidal proteins	Target insect
Corn	*Cry1Ab, Cry1F, Cry3Bb1 Cry1Ab, Cry1F, Cry2Ab, Cry9C, Vip3A*	European corn borer, fall armyworm, corn earworm
Cotton	*Cry1Ac, Cry2Ab*	Bollworm complex, armyworms
Soybean	*Cry1Ac*	Soybean looper, velvetbean caterpillar
Brinjal (Eggplant)	*Cry1Ac*	Eggplant fruit and shoot borer
Potato	*Cry3A*	Colorado potato beetle
Rice	*Cry1Ab, Cry1Ac, Cry2A*	Rice stem borer, leaf folder
Poplar	*Cry1Aa*	Leaf-eating lepidopteran pests
Sugarcane	*Cry1Ac, Cry2A*	Sugarcane borer
Tomato	*Cry1Ab*	Fruit borer
Apple	*Cry1Ab*	Lepidopteran pests (moths)
Cabbage	*Cry1Ac*	Lepidopteran pests (moths)

Initially, the *Bt* insecticidal-protein genes, *cry1Aa, cry1Ab,* and *cry1Ac* genes were not particularly well expressed in plants. Therefore, it was necessary to alter things in order to have an approach that would be commercially viable. Truncated versions of the *Cry* genes which included only the N-terminal region of the insecticidal protoxin (the portion of the protein with the insecticidal activity) were found to be much more effective. Further alterations of the protoxin gene included genes with codons altered so that only those codons that were more commonly used in plants were included. Tolerance to lepidopteran pests has been demonstrated in transgenic tomato plants expressing *Bt* toxins. In early studies, the resistant phenotype was expressed in the transformed plants that was attributable to the truncated form of a *Cry* gene placed under the transcriptional control of the *Cauliflower mosaic virus* 35S promoter. In this way, the greenhouse grown transgenic tomato plants showed higher tolerance to tobacco hornworms (*Manduca sexta*), tomato fruit worms (*Heliothis zea*), and tomato pinworms (*Keiferia lycopersicella*) than plants over-expressing the native form of the gene. Moreover, this insect resistance was reproducible in field-grown tomato plants.

As discussed earlier, a higher level of production of target proteins can be achieved when the genes are expressed in plant chloroplasts (Fig. 6.2). In one study, the insecticidal protein expression levels have been shown to increase to up to 45.3% of the total leaf protein when the *Bt cry2Aa2* gene operon was integrated into the tobacco chloroplast DNA. Insect pests such as cotton bollworm and beet armyworm were killed 100% after consuming these transgenic leaves. Furthermore, the protein was stably expressed at high levels in the chloroplast throughout the life cycle of the plant, potentially contributing to improved insecticidal efficacy. Moreover, the fact that chloroplasts are transmitted only through the ovule and not through the pollen, reduces the risk of unwanted transfer of the protoxin gene to other plants in the environment by pollen. Although, one drawback of expressing the *Bt* protoxin in chloroplasts is that insects that attack stems or fruit will not encounter the protoxin, since these tissues lack chloroplasts.

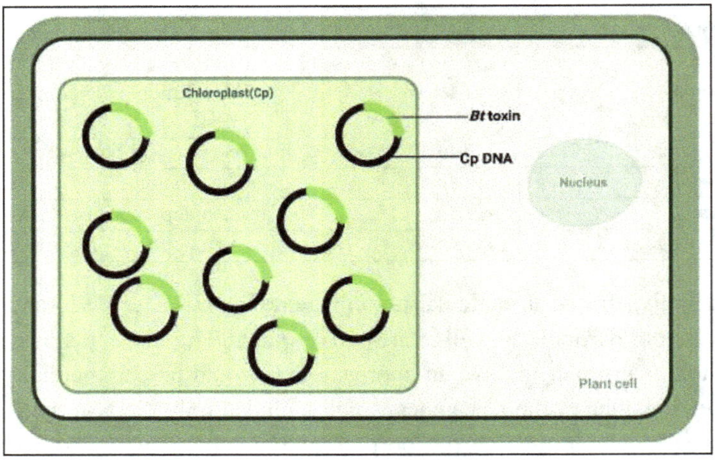

Fig. 6.2 A plant cell that has undergone plastid transformation showing the *Bt* gene integrated into the chloroplast genome

Transgenic plants deliver *Bt* toxins in a manner quite different from that of *Bt* microbial sprays. When *Bt* is applied as a foliar spray, its effectiveness is significantly limited by environmental factors such as light. Sunlight rapidly breaks down the Cry proteins, reducing their insecticidal activity within hours of application. This may result in a short window of *Bt* toxin effectiveness, requiring multiple rounds of spraying to maintain pest control, which increases labor and costs for farmers. By contrast, inserting *Bt* toxin genes directly into plants makes it possible to develop plants that are intrinsically resistant to target pests and do not need to be sprayed with costly and potentially hazardous chemical insecticides. It must be noted, however, that insect populations still have to be monitored as the development of resistance to the expressed *Bt* proteins is a possibility.

Vegetative insecticidal proteins (Vips)

To increase the target range of insect pests and to delay development of *Bt* resistance, the vegetative insecticidal proteins have been a tremendous complement to Cry proteins. It was discovered that clarified culture supernatant fluids collected during vegetative (i.e., log-phase) growth of *Bacillus* species are a rich source of Vips. The Vips have a mode of action distinct from the traditional *Bt* Cry toxins, and include several groups, each with its own insecticidal spectrum and efficacy. For example, groups Vip1 and Vip2, are not toxic to lepidoptera, whereas members of the Vip3 group target a wide spectrum of insects. This differential mode of action is exploited for increasing the insecticidal efficacy in plants and to combat the development of resistance to any single mechanism. To this end, scientists have performed domain shuffling i.e., combining parts of two Vip3 proteins, Vip3Ac1 and Vip3Aa1, into a single hybrid protein (Fig. 6.3). Swapping protein domains is a simple procedure in which analogous domains from different sources are swapped into a multi-domain protein to provide novel functionality. Many large proteins are composed of smaller independently folding domains which are often linked together by short peptide sequences. It is possible to swap domains from one protein to another to rebuild proteins and produce novel molecules. Of the two generated hybrids, one hybrid Vip3AcAa displayed enhanced toxicity to fall armyworms, cotton bollworms, and silkworms. The European corn borer (*Ostrinia nubilalis*) is a major pest of maize, while Vip3Ac on its own was not effective at killing this insect, transgenic plants expressing the chimeric Vip3AcAa exhibited a high level of resistance to this insect due to its combined structure. This means Vip3AcAa could potentially help control pests that Vip3Ac alone could not. Moreover, transgenic plants expressing this new chimeric protein were toxic to a Cry1Ac non-susceptible strain of cabbage looper.

Thus, by combining the chimeric toxin Vip3AcAa with Cry proteins in transgenic plants, pest control is enhanced through multiple modes of action, making it much more difficult for insect populations to develop resistance.

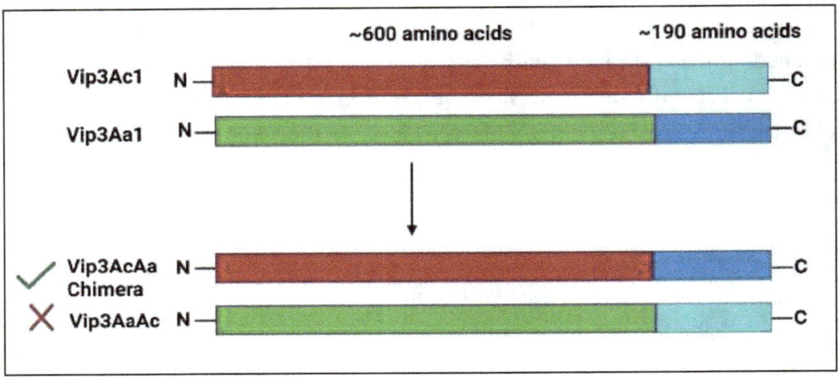

Fig. 6.3 Two new chimeric proteins generated by domain shuffling. The chimeric protein Vip3AcAa showed an enhanced insecticidal activity toward the insect larvae tested. The approximate sizes of the N- and C-terminal fragments of the proteins from Vip3Ac1 and Vip3Aa1 are indicated

RNAi interference

Double stranded RNA (dsRNA), the key molecule which triggers RNA interference (RNAi) (see Chapter 3), can be taken up by insects from transgenic plants via ingestion or microinjection. If the dsRNA is derived from a gene that encodes an essential insect function, it can silence that gene leading to insect mortality. For this strategy to be successful, the dsRNA must match the target gene sequence precisely. A large-scale RNAi screening experiment was conducted where the larvae of western corn rootworm (*Diabrotica virgifera*), one of the most destructive pests of corn, were fed with an artificial diet containing dsRNA. Feeding larvae with dsRNA targeting the insect midgut enzyme vacuolar ATPase (*v*-ATPase A) elicited a silencing response in those larvae, leading to significant insect mortality. *In planta* expression of this construct showed protection against root worm comparable to that provided by a *Bt* transgene. It is conceivable that this technology will be extended to more insect groups. However, it remains to be seen whether transgenic plants produced by this method show effective pest control under field conditions.

By identifying specific detoxification mechanisms that insects use, it is possible to target these pathways to manipulate resistance in plants. Cotton is capable of producing a yellow terpenoid (a naturally occurring organic compound derived from isoprene) called gossypol which protects the plant from insect predation. In insects, detoxification of this compound involves the enzyme cytochrome P450 monooxygenase. Thus, engineering plants to synthesize an RNAi molecule that silences the insect's gene for cytochrome P450 monooxygenase inhibits gene expression. Upon ingestion by the susceptible insect, the gossypol produced by the plant kills the insect and the damage to the transgenic plant is limited (Fig. 6.4).

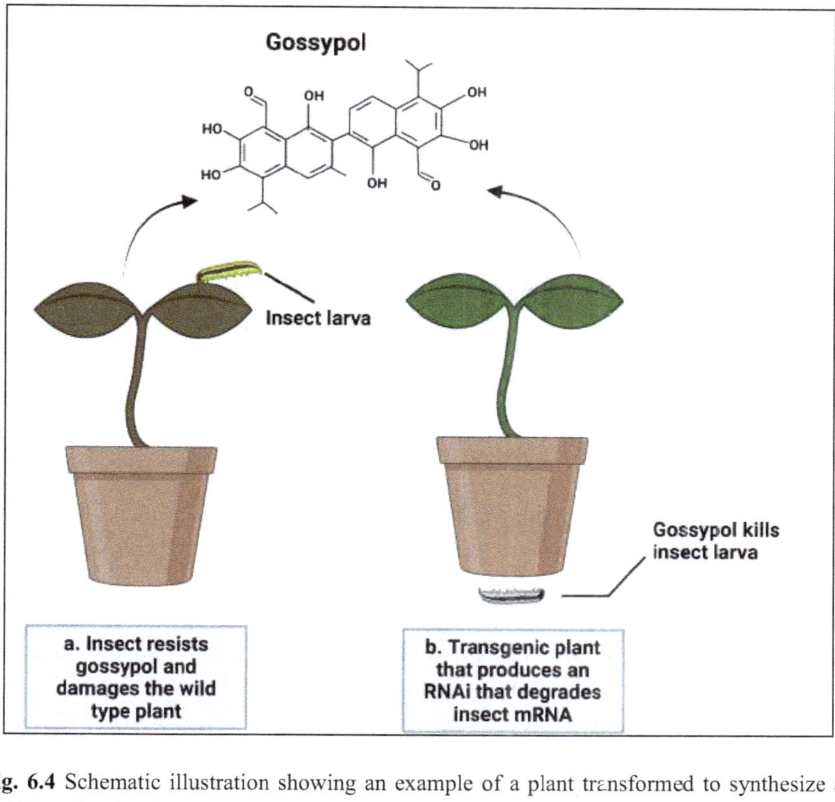

Fig. 6.4 Schematic illustration showing an example of a plant transformed to synthesize an RNAi molecule that silences the insect's gene for the cytochrome P450 monooxygenase causing the death of the insect larva (on the right side) whereas the non-transformed wild-type plant is severely damaged by the insect (on the left side). Also shown is the chemical structure of gossypol

6.2 Virus resistance

Viral diseases do not inherently cause plant death but often debilitate plants so that they become susceptible to infection by other pathogens. Many plant viruses are transmitted by vectors such as nematodes, aphids, whiteflies, or leafhoppers, which acquire the virus during the vector's feeding process. Plant viruses spread through the plant's vascular tissues to other parts of the plant to produce visible symptoms such as mosaic-mottle patterns and leaf curling to chlorosis (yellowing) of leaves to stunting of plant growth or death. The majority of plant viruses are RNA viruses, containing either one, two or three single-stranded plus-sense RNA as the genetic material. However, a smaller group of plant viruses carry DNA genomes, which can be either single stranded DNA or double stranded DNA (Table 6.2).

Table 6.2 Examples of some common plant viruses

Plant virus type	Genome type	Examples
RNA viruses	Plus-one stranded Plus-two stranded Plus-three stranded	Tobacco mosaic virus Cowpea mosaic virus Brome mosaic virus
DNA viruses	Single stranded	Cauliflower mosaic virus Bean golden mosaic virus Maize streak virus Beet curly top virus Cassava latent virus

Traditional approaches to combatting plant viruses including chemical treatments targeting the vectors and introducing naturally occurring virus resistance genes from related species or cultivars have had very little impact in controlling viral diseases, due to toxicity and resistance of viral vectors and lack of natural resistance genes in plants. Thus, other approaches to dealing with plant viruses have been developed. One method is to exploit the phenomenon of cross-protection, which is based on earlier studies that infection of a plant initially with a virus confers protection to subsequent infection with a virulent strain of the same or closely related virus that normally causes severe symptoms. This phenomenon is related to the expression of viral coat protein genes and has been extensively used to protect transgenic plants against viral diseases.

Viral coat protein mediated protection

Upon entry into a plant cell, the virus particle sheds its protein coat (which is usually the most abundant protein of the virus particle) to release its DNA or RNA for replication. When a plant is genetically engineered to express the viral coat protein gene it may disrupt the synthesis of virus coat protein and genome replication. Consequently, the virus's capacity to produce progeny particles is hampered. Transgenic coat protein resistance can be used to control many important virus diseases of crops (Table 6.3). A successful example of this strategy is the engineering of papaya (*Carica papaya)* plants against papaya ringspot virus (PRSV). In the 1950s, the PRSV was discovered to be present on the Hawaiian island of Oahu and this devastating aphid-transmitted viral disease nearly wiped out the Hawaiian papaya industry. However, expressing the coat protein gene of PRSV in transgenic papaya dramatically reduced virus infestation and significantly increased yields and fruit quality in two cultivars "SunUp" and "Rainbow" of papaya. Resistance to PSRV in transgenic plants has held up under field conditions in Hawaii, without any adverse effects on human health, thus saving the Hawaiian papaya industry, when other methods of PRSV control failed.

Table 6.3 List of crops that have been developed based on the viral coat protein gene approach

Viral source of coat protein	Transgenic crop	Status
Papaya ringspot virus	Papaya ("Rainbow" and "SunUp" cultivars)	Commercially approved
Zucchini yellow mosaic virus	Squash (Zucchini)	Commercially approved
Watermelon mosaic virus	Squash (Zucchini)	Experimental
Cucumber mosaic virus	Cucumber	Experimental
Potato leafroll virus	Potato	Experimental
Potato virus Y	Potato	Experimental
Tobacco mosaic virus	Tobacco	Experimental
Tomato mosaic virus	Tomato	Experimental
Cucumber green mottle mosaic virus	Melon, watermelon	Experimental
Banana bunchy top virus	Banana	Experimental
Rice tungro bacilliform virus	Rice	Experimental
African cassava mosaic virus	Cassava	Experimental
Bean golden mosaic virus	Bean	Experimental
Cymbidium ringspot virus	Tobacco	Experimental
Grapevine chrome mosaic virus	Tobacco	Experimental
Maize dwarf mosaic virus	Sweet corn	Experimental
Beet necrotic yellow vein virus	Sugar beet	Experimental
Alfalfa mosaic virus	Tobacco, tomato, alfalfa	Experimental

Virus resistance in plants can also be accomplished by the expression of viral sequences in the antisense orientation. An antisense RNA molecule that is complementary to a particular mRNA will base-pair with the mRNA and prevent the mRNA from being translated. When a plant becomes infected with an actual virus, the antisense RNA binds to the viral mRNA resulting in a duplex RNA, thereby preventing the mRNA from being translated, thus repressing the corresponding mRNA in the cell. This prompted scientists to construct transgenic plants that synthesize antisense copies of viral coat protein genes. Subsequently, plants were challenged with viruses and exhibited resistance at lower concentrations of the virus.

It is often advantageous to engineer plants to be resistant to more than one virus. Considering this, Ti plasmid binary vectors expressing multiple coat protein genes for cucumber mosaic virus (CuMV), zucchini yellow mosaic virus (ZYMV) and watermelon mosaic virus 2 (WMV2) were used to transform yellow crookneck squash (Fig. 6.5). Transgenic plants that contained the coat protein genes from all three viruses were resistant to damage by all three viruses under laboratory conditions. Moreover, transgenic plants expressing coat proteins for ZYMV and WMV2 were tested for the capacity to suppress disease symptoms under field conditions. The transgenic plants that expressed both coat protein genes were completely resistant to infection by these viruses, a result that indicates that it may be possible to control multiple infections under field conditions.

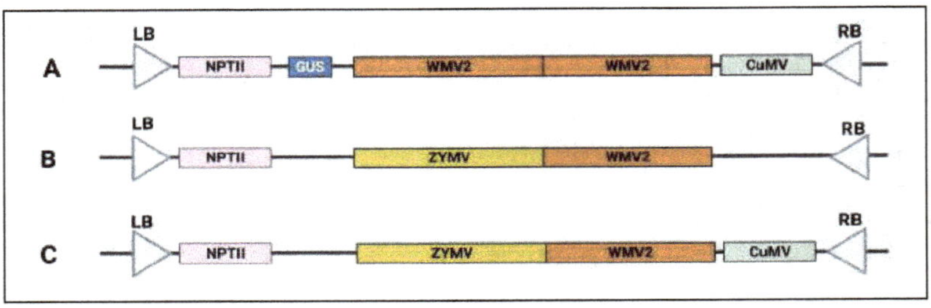

Fig. 6.5 T-DNA construct of the binary plasmid vectors used for the genetic transformation of yellow crookneck squash (A) neomycin phosphotransferase (NPTII) gene as a selectable marker, a β-glucuronidase (GUS) gene as a reporter gene, RB right border, LB left border, two copies of the coat protein gene from watermelon mosaic virus 2 (WMV2), and the coat protein gene from CuMV (B) similar to part A without CuMV and GUS, with one copy of WMV2, and with the coat protein gene from ZYMV (C) same as part B, with the addition of CuMV. DNA constructs are engineered to contain both regulatory promoter elements and transcriptional terminator sequences, all of which are not shown

Other experiments on genetic engineering for plant virus resistance have highlighted a role for micro-RNAs (miRNAs), which are derived from noncoding RNA transcripts that fold back on themselves. This characteristic structure is then processed by Dicer like enzymes to produce mature miRNA, a short dsRNA which is then loaded into the RNA-induced silencing complex (see Chapter 3). The miRNA-RISC complex then interferes with viral replication by targeting the viral mRNA for degradation. It is possible to change the sequence of the 21-nucleotide long mature miRNA and produce artificial miRNAs in plants to target new viral genes. For example, it has been shown for turnip yellow mosaic virus (TYMV) and turnip mosaic virus (TuMV) that transformation of *A. thaliana* with a chimeric gene construct containing artificial miRNA sequences produces plants that exhibit resistance to the development of symptoms following infection with TYMV and TuMV. This approach offers another opportunity for developing plants that are resistant to viruses.

Instead of genetically modifying a plant to express dsRNA internally (as in RNAi transgenic crops), scientists have shown that they can successfully apply synthetic dsRNA molecules exogenously, which involves spraying the dsRNA onto plant surfaces. The dsRNA that is taken up by the plant induces its antiviral RNAi response (Fig. 6.6). This technique has been used effectively against the pepper mild mottle virus, alfalfa mosaic virus, and tobacco etch virus.

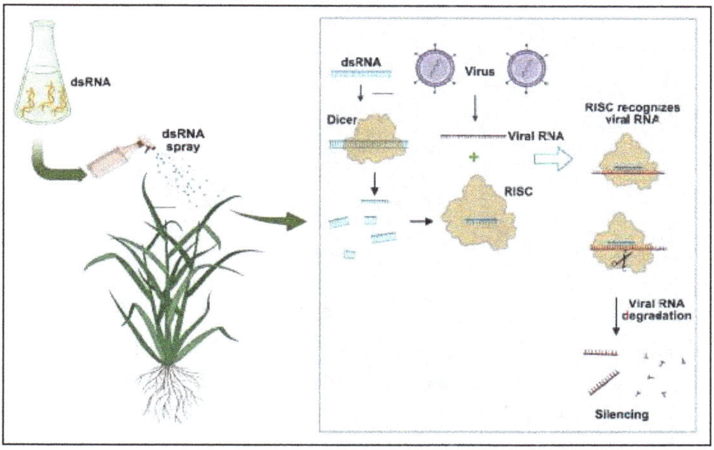

Fig. 6.6 Topical application of dsRNA. When dsRNA molecules are applied to plant foliage surfaces, they may be absorbed and processed by the plant's RNAi machinery. This processing leads to the degradation of complementary viral RNA sequences, effectively silencing the targeted viral genes and preventing the virus from replicating within the plant cells

Geminiviruses cause considerable damage to several families of monocotyledonous and dicotyledonous plants. Based on their insect vector they are classified into two major subgroups, i.e., whitefly transmitted (infect monocots) or leaf hopper transmitted (infect dicots) viruses. Their virion includes a twinned capsid structure and a relatively small circular genome of single stranded DNA. Gemini virus infection initiates with the release of ssDNA into the plant cell nucleus, which is converted to an intermediate dsDNA, from which new ssDNA is generated by rolling-circle replication (Fig. 6.7). The CRISPR/Cas9 system has been effectively utilized to engineer plants that target and cleave the viral RNA transcripts leading to the degradation of the viral genome, resulting in loss of susceptibility to viruses. After demonstrating viral resistance in the model plants *Arabidopsis* and tobacco, the same system was subsequently used in barley against wheat dwarf virus. However, it is likely that extended expression of the Cas9 nuclease may result in off target effects, leading to unwanted mutations in the host genome. Therefore, strict expression control should be used so that the Cas9 anti-viral system will only be expressed when the virus invades the plant cell. One possibility is the use of virus-induced promoters instead of a constitutive promoter.

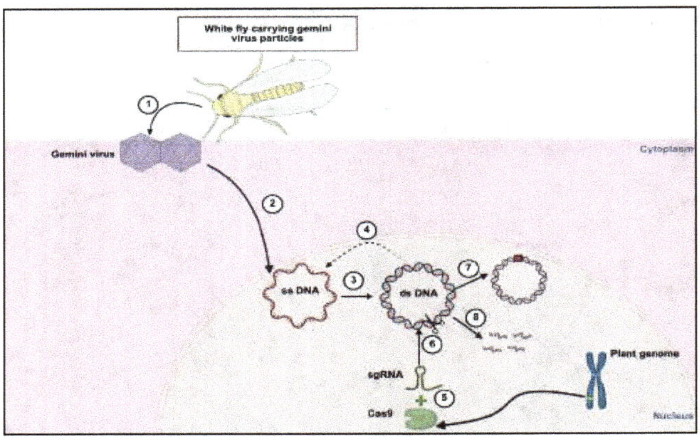

Fig. 6.7 Schematic diagram of the CRISPR/Cas9 mechanism that is used to target geminiviruses. 1. A white fly transmitting the geminivirus 2. The virus ssDNA enters the plant cell's nucleus. 3. The virus ssDNA is converted to a dsDNA intermediate structure 4. Additional copies of ssDNA are synthesized by rolling circle replication 5. The Cas9/sgRNA originating from the plant genomic DNA is synthesized within the plant cell's nucleus 6. The CRISPR/Cas9 complex binds to and cleaves the target sequence in the double stranded viral replicative form. Direct targeting of the viral genome sequence can lead to mutation or degradation of the viral DNA (steps 7 & 8). Note that for illustrative purposes, the insect and the geminivirus are not drawn to scale

A successful approach has been followed for engineering plant RNA potyviruses which are generally transmitted by aphids (Fig. 6.8). These viruses are filamentous, non-enveloped and rely on plant host initiation factor 4E (eIF4E) for protein translation. Mutant forms of eIF4E are an important source of resistance in plants against viruses that depend on them for replication. For example, transgenic cucumber plants containing mutated *eIF4E* genes have been produced by using CRISPR/Cas9. These plants show broad spectrum resistance to cucumber vein yellowing virus, papaya ringspot virus, and zucchini yellow mosaic virus. A similar phenotype was observed for gene edited cassava plants, where the plants had some level of resistance to casava brown streak virus. However, developing eIF4E-based resistance may sometimes lead to impaired plant growth. This problem can be fixed by introducing point mutations in the gene (instead of knocking it out), which does not impact the normal physiological process in the plant but prevents interaction with the virus.

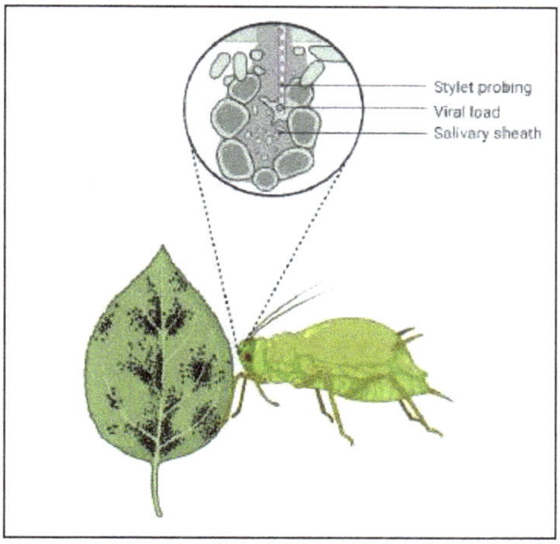

Fig. 6.8 Schematic representation of an aphid-mediated viral transmission in plants. Aphids quickly acquire the virus from an infected plant and spread it to a non-infected plant inflicting severe damage. Note that for illustrative purposes, the insect and the plant leaf are not drawn to scale

The CRISPR/Cas system has been successfully employed to target numerous viruses in many plant species (Table 6.4). Furthermore, the advantage of expressing numerous gRNAs with a single Cas protein has allowed scientists to target multiple DNA viruses in transgenic plants, pointing to the feasibility of controlling multiple viruses with a single engineered system. For instance, simultaneously targeting two sites of the viral genome was more highly effective in interfering with proliferation of chili leaf curl virus, than individually targeting a single site. As a further development of this method, scientists were able to engineer plants with CRISPR/Cas9 to simultaneously modify six genes of the cotton leaf curl virus. It remains to be determined whether this resistance is effective and durable over the long term, particularly under field conditions.

Table 6.4 Examples of CRISPR/Cas9-based virus resistance in plants

Host plant	Resistant to virus	Genome type/target gene	Viruses genus
Arabidopsis thaliana	Beet severe curly top virus Turnip mosaic virus Clover yellow vein virus	ssDNA *elf4E* gene *elF4E1*gene	Geminivirus Potyvirus Potyvirus
Nicotiana benthamiana	Cotton Leaf Curl Multan Virus	ssDNA	Geminivirus
Hordeum vulgare	Wheat dwarf virus	ssDNA	Geminivirus
Triticum aestivum L.	Wheat dwarf virus	ssDNA	Geminivirus
Manihot esculenta	African cassava mosaic virus	ssDNA	Geminivirus
Solanum lycopersicum	Tomato yellow leaf curl virus Pepper mottle virus Potato virus	ssDNA *Elf4E* gene *Elf4E* gene	Geminivirus Potyvirus Potyvirus
Cucumis sativus L.	Zucchini yellow mosaic virus Papaya ring spot mosaic virus	*Elf4E* gene	Potyvirus Potyvirus

6.3 Herbicide resistance

Weeds are a serious problem in farmlands worldwide because they compete with crops for space, sunlight, water and fertilizer, causing extensive reductions in crop yields. Chemical herbicides are widely used by farmers to eliminate weeds. However, such compounds have toxicity to animals and soil organisms and many of them kill plants non-selectively. Additionally, the full amount of the applied herbicide does not always reach the target site, and many herbicides leach into groundwater or run off into surface water.

Herbicides act by targeting specific biological processes in plants. Some types of herbicides kill plants by inhibiting photosynthesis; however, these do not discriminate weeds from crop plants and are harmful to animals. There are other types of herbicides which kill plants by inhibiting the synthesis of amino acids. Animals require amino acids in their diet, whereas plants and bacteria have to synthesize them. Therefore, such herbicides are not harmful to animals. Because these herbicides are non-selective, most plants sprayed would be killed, thereby limiting their usefulness. With the advent of transgenic technology, a large amount of effort is being made to develop crops that are genetically modified so that they can tolerate high levels of herbicides.

Glyphosate (N-(phosphonomethyl)glycine) is the most widely used non-selective herbicide. Glyphosate inhibits 5-enolpyruvylshikimate-3-phosphate synthase (EPSP), a key enzyme in the shikimic acid pathway, which is essential in the synthesis of phenylalanine, tyrosine and tryptophan (Fig. 6.9).

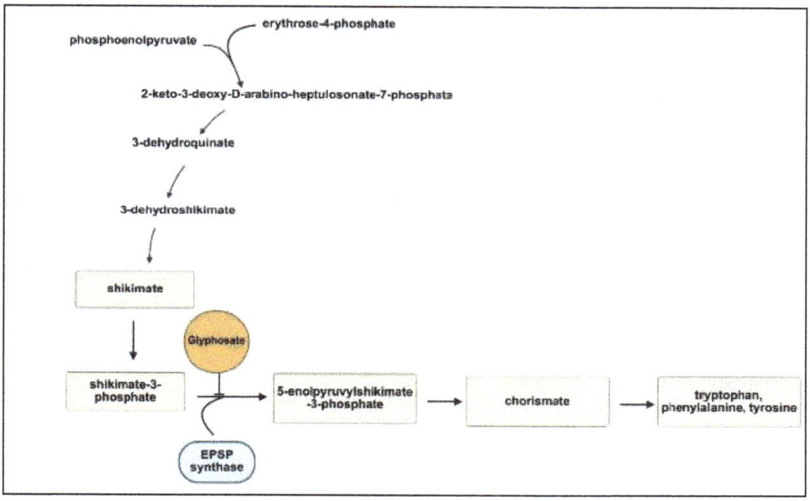

Fig. 6.9 The shikimate pathway is important for the synthesis of essential amino acids in plants and bacteria. Glyphosate inhibits the reaction catalyzed by EPSP synthase enzyme, which prevents the production of the amino acids, tryptophan, tyrosine and phenylalanine in cells. Redrawn from Kougan et al. (2013)

Trademarked as Roundup, and developed by the Monsanto Corporation, glyphosate has low toxicity to animals and has a short life in soils. Moreover, it is highly active to a wide range of weeds. Genes encoding EPSP synthase have been isolated from bacteria and have been used to engineer glyphosate resistance in plants. To develop "Round-up ready" plants, scientists transferred the EPSP synthase gene from a glyphosate-resistant strain of *E. coli* and expressed it in transgenic plants. The bacterial sequence and appropriate plant transcription initiation and termination signals were incorporated into transgenic plants via Ti plasmids. Transgenic plants show resistance to glyphosate because the sensitive target for the herbicide is replaced by a resistant one. Transgenic corn, canola, tomato, potato, sorghum, bean, sugar beet and cotton plants developed by this approach were able to achieve commercial-level tolerance to glyphosate. Unfortunately, the use of Roundup has become controversial. Recently, the World Health organization has classified Roundup as probably carcinogenic to humans.

Rice is a major cereal crop for more than half of the world's population. But rice production is severely affected by weeds, often causing yield losses ranging from 15% to 100%, especially in direct seeded systems. Although the advantages of glyphosate are obvious, it is not applied directly in rice fields due to its toxicity to rice. It is known that glyphosate inhibits the native plant EPSPS enzyme, however the native gene of EPSP synthase that is isolated from the actinomycete species, *Isoptericola variabilis* is not sensitive to glyphosate. To ensure that this gene would be expressed at a high level in transgenic rice plants, the gene was modified to reflect the codon usage that is found in rice plants. Following introduction of the codon-optimized *I. variabilis epsp* synthase gene into rice plants by *Agrobacterium tumefaciens,* the *epsps* gene was strongly expressed under the control of a maize ubiquitin promoter, which is highly active in monocots like rice. As a result of these manipulations, the regenerated rice plants had a high level of glyphosate tolerance (Fig. 6.10). Thus, rice plants engineered by this approach have the potential to be used in rice fields for weed control with glyphosate.

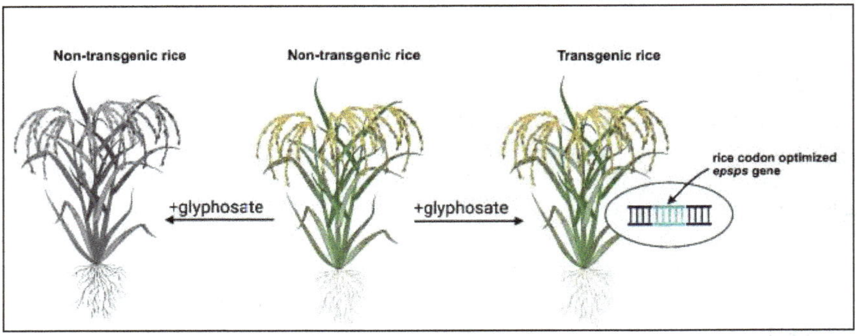

Fig. 6.10 Generation of glyphosate-resistant rice plants by introducing a codon-modified *I. variabilis epsps* gene. The phytotoxic effect typically caused by the herbicide glyphosate is also shown

Another approach that has been worked out for the engineering of glyphosate resistance in plants is introducing bacterial detoxifying enzymes. In this case, researchers first isolated a glyphosate N-acetyl transferase (*gat*) gene from *Bacillus licheniformis*. This gene encodes an enzyme that converts glyphosate into nontoxic N-acetylglyphosate. In the next step, the *gat* genes were functionally improved by gene fragment shuffling, to optimize the enzyme catalytic efficiency (the k_{cat}/K_m ratio). Finally, the modified *gat* gene was introduced into *Arabidopsis,* tobacco, and corn plants. The transgenic plants expressing the modified *gat* gene show a six-fold resistance to glyphosate compared with the non-transgenic plants.

Sometimes multiple genes can work together to enhance a trait more effectively than a single gene. For example, it was shown that the combination of two herbicide-tolerance genes with different mechanisms show superior tolerance to glyphosate than plants engineered with one gene. Thus, transgenic corn plants co-expressing *gat* and *epsp* synthase genes have been produced by transformation with an *Agrobacterium* Ti plasmid vector. Under field conditions, the transgenic plants were resistant to ten times the normal killing dose of glyphosate.

In general, continuous use of any one herbicide generally leads to the evolution of weeds resistant to that herbicide. Intensive use of glyphosate has resulted in the development of resistance to these herbicides in 32 different weeds globally. As a result, it is envisaged that crops that are resistant to multiple herbicides will be required to slow down the development of resistant weeds. In this regard, dicamba is a widely used, environmentally friendly herbicide which acts by mimicking the plant growth hormone indole-3-acetic acid. No weed species are reported to be resistant to dicamba. In order to develop transgenic plants resistant to dicamba, a bacterial dicamba monooxygenase (DMO) gene was isolated and subsequently expressed in *Arabidopsis,* tomato, and tobacco plants. This gene from the bacterium *Pseudomonas maltophilia* provides a mechanism of herbicide resistance by demethylating the methoxy group of dicamba and converting it to a hydroxyl group therefore rendering it inactive against dicamba. As a result, transgenic plants show tolerance to high doses of dicamba when grown under greenhouse and field conditions. Moreover, it is possible to engineer a single plant with genes encoding both dicamba and glyphosate resistance. A weed is unlikely to develop resistance to both herbicides, while the doubly transformed plants remain unaffected by these herbicides.

Similarly, resistance to another important non-selective herbicide glufosinate has been achieved by transferring a bacterial gene encoding a detoxifying enzyme. Glufosinate inhibits glutamine synthase (GS), a key enzyme in the nitrogen assimilation pathway, detoxifying ammonia, and supplying glutamine for biosynthesis of nucleotides, amino acids, and other nitrogenous compounds. Accumulation of toxic ammonia due to inhibition of GS by glufosinate results in plant death. The *bar* gene isolated from the bacterium *Streptomyces hygroscopicus* encodes the enzyme phosphinothricin acetyltransferase (PAT) which detoxifies glufosinate by acetylating the free NH_2 groups (Fig. 6.11). The *bar* gene was subcloned, put under the control of the 35S promoter from *Cauliflower mosaic virus*, and expressed in tomato, tobacco, and potato plants. Under greenhouse conditions, these transgenic plants demonstrated significant resistance to the glufosinate herbicide. Many commercially important glufosinate-resistant crops such as canola, corn, cotton and soybean have been developed by introducing *bar* genes and these transgenic crops have been successfully cultivated in many parts of the world.

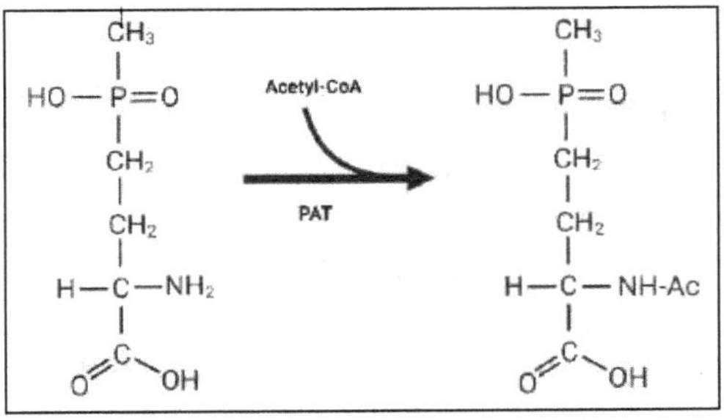

Fig. 6.11 *N*-acetylation of the herbicide glufosinate by the bacterial enzyme phosphinothricin acetyltransferase (PAT)

Apart from generating transgenic crops tolerant to various herbicides, scientists have also been able to develop herbicide tolerance in some crop plants with the help of CRISPR/Cas9. Acetolactate synthase (ALS) is a key enzyme which catalyzes the initial step in the biosynthesis of the branched-chain amino acids (leucine, isoleucine and valine) and is the target of many commercially used herbicides including sulfonylureas and imidazolinones. This enzyme is encoded by nuclear genes that contain a transit peptide sequence for transport of the pre-protein into the chloroplast. Herbicide binding appears at a specific site on the ALS enzyme. Since the biochemical site of action of this herbicide is known, researchers have used CRISPR/Cas9 to introduce site specific mutations in the

ALS gene which resulted in producing an altered enzyme that was no longer inhibited by such herbicides. Accordingly, many crops such as rice, maize, soybean, wheat, watermelon, canola, tobacco, tomato and potato have been successfully transformed using the CRISPR/Cas9 method to introduce similar mutations, enhancing their tolerance to ALS-inhibiting herbicides (Fig. 6.12).

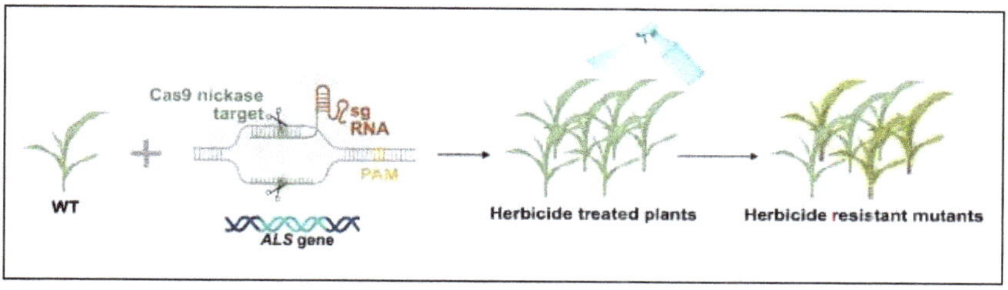

Fig. 6.12 Schematic view of crops resistant to ALS-inhibiting herbicide developed by CRISPR/Cas9 technology

Often, soils contaminated with herbicides harbor microorganisms capable of degrading herbicides. To find bacteria that can degrade bromoxynil (3,5-dibromo-4-hydroxybenzonitrile), a photosynthesis inhibitor, one group of researchers developed a selective screening strategy (Fig. 6.13). The test was based on the ability to measure the growth of cells and the amount of bromoxynil remaining in the culture filtrate by high performance liquid chromatography (HPLC). One isolate, identified as *Klebsiella ozaenae*, could metabolize bromoxynil and use it as the sole source of nitrogen. Further analysis suggested the role of a nitrilase enzyme that acts on the cyano group of bromoxynil and inactivates it before the herbicide can act. To engineer bromoxynil-resistant plants, researchers took advantage of this enzyme and placed the bacterial gene under the control of the Rubisco small sub-unit light responsive element before it was transferred to tobacco and tomato plants. As expected, the transgenic plants expressed nitrilase activity in their shoots and leaves, but not in their roots, and were resistant to the toxic effects of the herbicide.

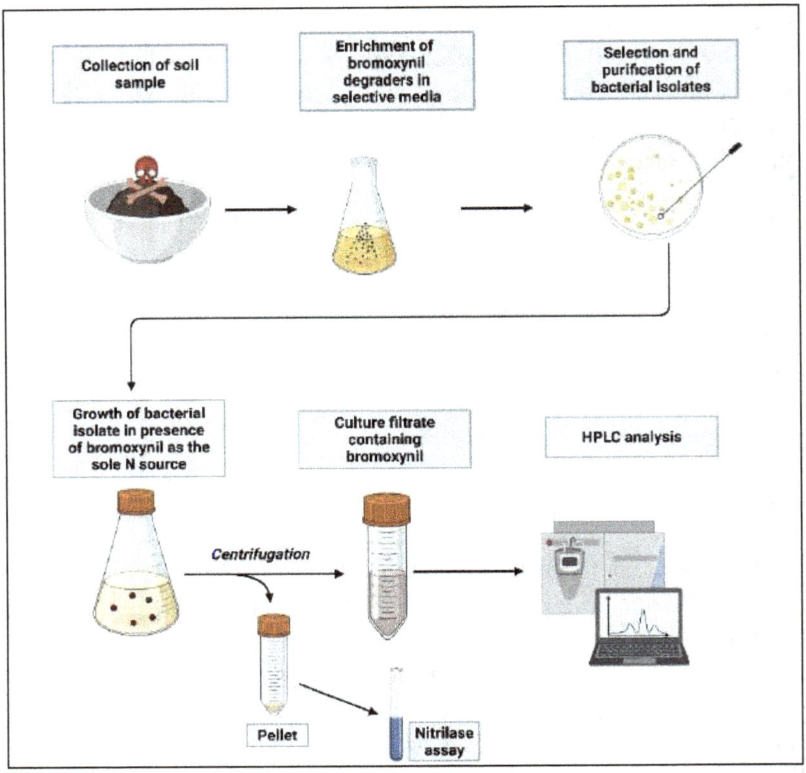

Fig. 6.13 Overview of a scheme used to isolate bromoxynil degrading bacteria from contaminated soils

6.4 Bacterial and fungal resistance

Phytopathogens including fungi and bacteria reduce the yield and quality of agricultural production causing substantial economic losses globally. Plants react to bacterial or fungal pathogen invasion by activating a series of defense mechanisms. The synthesis of salicylic acid appears to be necessary for the activation of plant defense genes i.e., the pathogenesis-related (PR) proteins, which lead to the induction of systemic acquired resistance in the plant (Fig. 6.14). Scientists have attempted to manipulate these components of systemic acquired resistance to engineer disease resistance in plants.

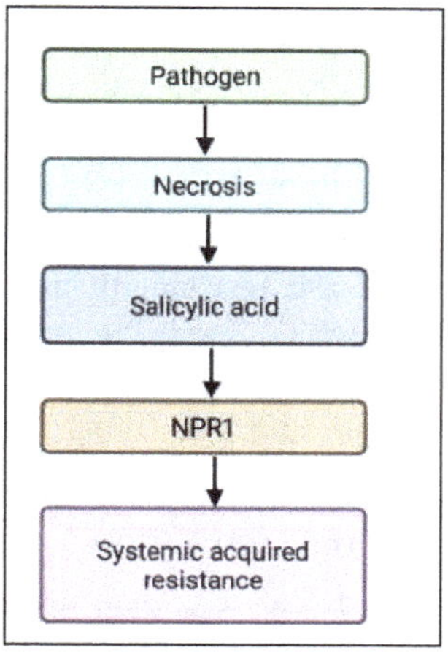

Fig. 6.14 An overview of systematic acquired resistance

In the NPR (Nonexpressor of PR genes) protein family, the NPR1 gene, plays a central role in mediating the plant's response to SA. Therefore, overexpression of NPR1 in crop plants has been widely used to enhance resistance against a variety of fungal pathogens. For instance, increased resistance to a variety of fungal infections was obtained by the overexpression of the *Arabidopsis thaliana NPR1* gene in strawberry plants (Fig. 6.15).

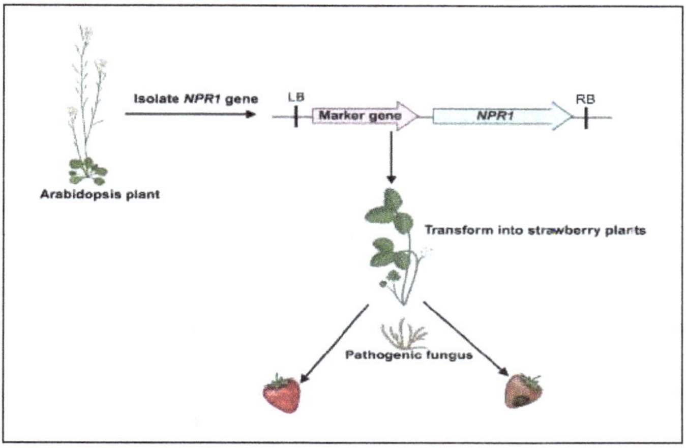

Fig. 6.15 Schematic representation of the generation of transgenic strawberry plants that overexpress an *NPR1* gene from *A. thaliana*. Following screening, only those transformants (the strawberry on the left) that produce fungus-resistant transgenic plants are selected for further propagation

PR proteins are key players in the plant's systemic acquired resistance. Many studies have shown that PR proteins can hydrolyze the fungal cell wall components, resulting in cell death. Researchers have attempted to target many members of the PR proteins family including PR2 (e.g., β-1,3-glucanases), PR3 (e.g., chitinases), PR5 (thaumatin-like proteins) and PR-6 (protease inhibitors) for generating broad spectrum disease resistant plants against fungal pathogens. Especially, there are many reports showing that over expression of plant chitinase genes confers broad-spectrum disease resistance in plants modified by genetic engineering. Transgenic plants including rice, tobacco, and canola that overexpressed chitinase under a strong *cauliflower mosaic virus* 35S promoter are resistant to the fungal pathogen (Fig. 6.16).

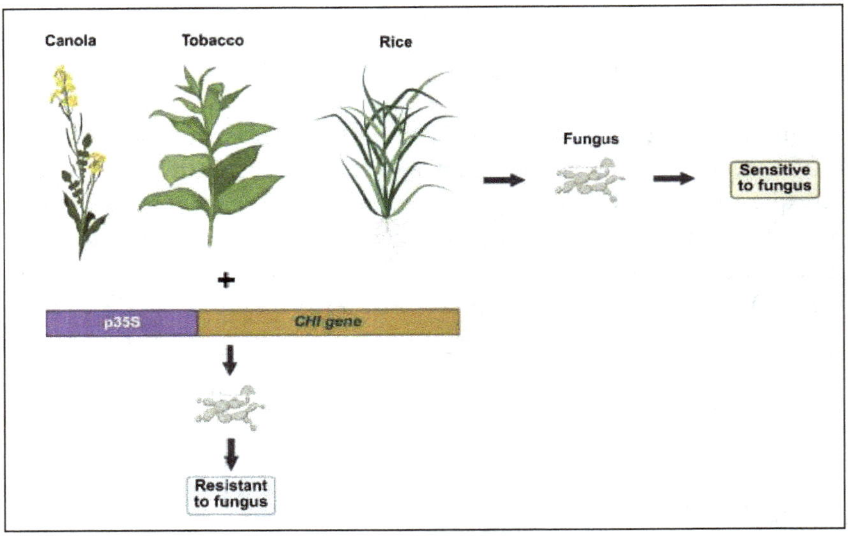

Fig. 6.16 Canola, tobacco and rice plants that have been genetically transformed to express the *CHI* gene (from rice) are resistant to the fungal pathogen (*Magnoporthe grisea)* that causes rice blast. The protein product of the *CHI* gene can hydrolyze the β-1,4 linkages of the *N*-acetyl-d-glucosamine polymer chitin, a major component of many fungal cell walls

Furthermore, the use of heterologous chitinase genes has led to the development of transgenic plants with increased tolerance to chitin-containing (in their cell walls) fungal pathogens. Bitter melon (*Momordica charantia*), which has high antifungal activity in leaves, is highly tolerant to various pathogens. Transgenic rice plants overexpressing the *McCHIT1* chitinase gene from melon, showed moderate levels of resistance to *Magnaporthe grisea* (rice blast) and

Rhizoctonia solani (sheath blight), two major fungal pathogens of rice Although, the above result offers an efficient target to confer disease resistance, fine tuning of engineered gene expression may be necessary in future research.

Fusarium head blight (FHB) which is also known as head scab and ear blight, caused by *Fusarium graminearum* Schwabe, is one of the most devastating wheat diseases worldwide. Every year, infection of wheat with this pathogen leads to crop losses valued at approximately $5.6 billion. The United States Department of Agriculture (USDA) ranks FHB as the worst plant disease to hit the US since the rust epidemics in the 1950s. This pathogen reduces plant yields by producing the toxin deoxynivalenol (DON; also called vomitoxin), that renders the grain too toxic to use. While the Fusarium head blight resistance gene, *Fhb1,* is present in modern wheat cultivars, it often exhibits partial resistance to the disease. The recent discovery of the *Fhb7* gene from a distant relative of wheat provides a valuable resource for developing FHB resistant plants. Originally transferred from a fungal endophyte (*Epichloe* sp.) into tall wheat grass (*Thinopyrum ponticum*), *Fhb7* has been successfully introduced into a strain of wheat using transgenic methods. Unlike the *Fhb1* gene that blocks the fungal spread, *Fhb7* encodes a glutathione S-transferase enzyme that detoxifies deoxynivalenol by conjugating it to glutathione, preventing the growth and spread of the pathogen in the infected plant.

A lot of effort has also been devoted to the creation of transgenic resistance to Chestnut blight, a fungal disease caused by *Cryphonectria parasitica,* which wiped out an estimated four billion American Chestnut trees. One of the major virulence factors of *C. parasitica* is the production of oxalic acid, which is toxic to chestnut trees because it lowers the pH of the infected plant tissues from around the normal pH 5.5 to around 2.8, killing plant cells (Fig. 6.17). Many plants produce oxalate oxidase (*oxo*), which detoxifies oxalic acid and confers resistance against pathogens that use that strategy. Based on this, transgenic chestnut trees that express a wheat oxalate oxidase gene (*oxo*) were developed. Plants from transformed cell lines and their sexual progeny produce the enzyme and have enhanced resistance to the disease. It is anticipated that, through the use of this approach, chestnut trees will return to Eastern U.S. forests.

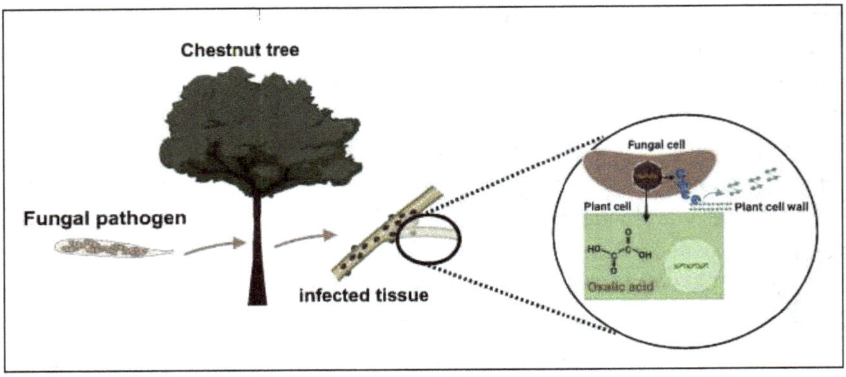

Fig. 6.17 The fungal pathogen infects the chestnut tree and produces oxalic acid, which contributes to host tissue damage due to the increase in pH

Plant bacterial pathogens such as *Pseudomonas syringae*, *Xanthomonas* spp. and *Erwinia* spp. often secrete phytotoxins or virulence factors which allow the pathogen to multiply in the infected plant tissue. Transgenic plants have been produced with genes that encode enzymes capable of detoxifying several bacterial toxins. For example, tabtoxin, produced by *P. syringae* pv. *tabaci* inhibits glutamine synthetases and causes toxic ammonia accumulation resulting in chlorotic halos on leaves. In order to increase disease resistance against this pathogen, an acetyl transferase gene that encodes an enzyme that detoxifies tabtoxin was introduced into plants. Similarly, an ornithine carbamoyltransferase gene was introduced into plants to detoxify a phaseolotoxin produced by *P. syringae* pv. *phaseolicola.* Northern corn leaf spot, caused by *Cochliobolus carbonum* produces HC-toxin, a cyclic tetrapeptide. Scientists engineered maize to express a detoxifying enzyme HC-toxin reductase, resulting in resistant transgenic maize. These strategies have rendered transgenic plants resistant to these toxin-producing pathogens. In cases where the gene is known, using any of the methods described above is straightforward. However, this approach is effective only against the limited number of pathogens that are known to cause disease by producing phytotoxins. Alternatively, genes that produce anti-microbial proteins (AMPs) are less specific and may be a solution to this problem. Typically, AMPs act by disrupting the microbial cell membrane, for instance using cecropin, a lytic peptide that targets the conserved pathogen lipid molecules creating holes in the bacterial outer membrane (Fig. 6.18). Transgenic apples, potato, and tobacco with genes for these lytic peptides have been generated, which confer some protection against bacterial wilt and fire blight caused by *Erwinia amylovora.* Another potential method for providing protection against bacterial pathogens is by engineering fusion proteins containing cecropins, which are translated as a single polypeptide and subsequently cleaved to their active (smaller) constituents that target and destroy the phytopathogenic bacteria.

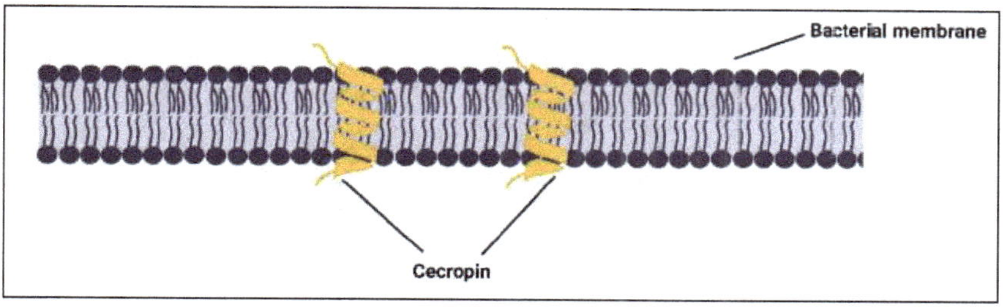

Fig. 6.18 A schematic drawing of cecropins inserted into the bacterial cell outer membrane

This strategy was employed to engineer resistance to *Xylella fastidiosa*, a bacterium affecting grapevines (the causative agent of Pierce's disease), olive (causing olive quick decline syndrome), and citrus (the cause of citrus variegated chlorosis) plants. The DNA construct, consisting of a fusion protein combining an elastase which binds to the MopB protein, an outer membrane protein of *X. fastidiosa,* with cecropin, was expressed in grape plants. Both the elastase gene (which digests the bacterial outer membrane) and the cecropin gene (which inserts into the lipid bilayer forming pores) work synergistically reducing pathogen-induced damage by up to 100-fold compared to using either molecule alone. The fusion protein was placed under the control of a plant-specific signal sequence, which directs the secretion of the fusion protein to the xylem, where the invading bacteria reside (Fig. 6.19). Thus, by identifying and targeting outer membrane proteins unique to other pathogens, similar fusion proteins can be engineered for broader pathogen coverage.

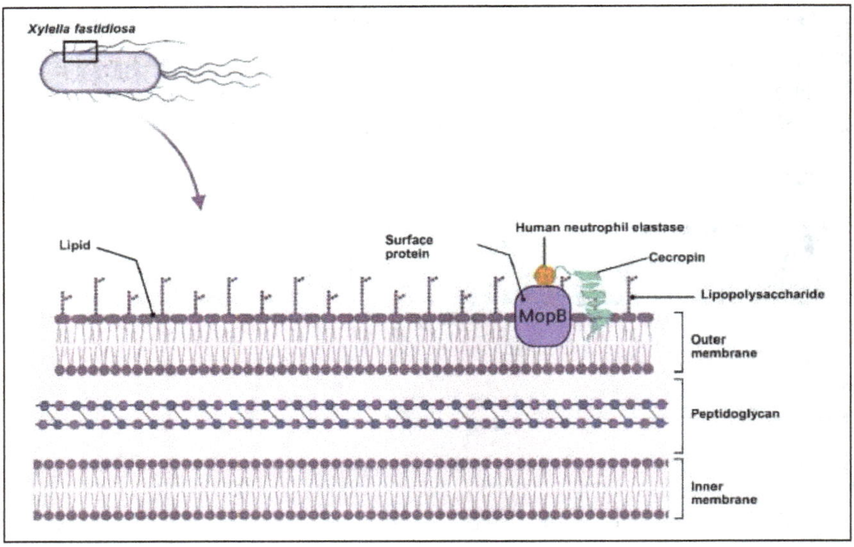

Fig. 6.19 Schematic drawing of a Gram-negative bacterial cell wall. A fusion protein consisting of two molecules-the protease human neutrophil elastase (the portion that binds to the bacterial outer membrane protein) and a lytic peptide (the portion that inserts into the lipopolysaccharide rich outer membrane and creates pores) was designed and used to create transgenic grape plants expressing this protein

6.5 Tolerance to salt and drought stress

Salinity and drought stress represent distinct types of abiotic stress, each arising from a different environmental cause, however, physiologically they share similar consequences for plants such as osmotic stress and oxidative damage. Thus, genetic engineering of plants to enhance salt and drought tolerance has been achieved by targeting one or more of these mechanisms, however, some strategies are specific for one stress or the other.

Engineering ion transport components

In halophytes (salt tolerant plants), V-ATPase, a multi enzyme unit is involved in resistance to salt stress, representing a potential target to enhance plant salt tolerance. The tolerance to salt in these plants is due to vacuolar sequestration of Na^+, mediated by the Na^+/H^+ antiport protein encoded by the V-ATPase gene (Fig. 6.20). The Na^+/H^+ antiport protein relies on the proton gradient (generated by the vacuolar H^+-translocating enzymes) to transport Na^+ ions into the plant vacuole, thereby preventing toxic sodium build-up. Therefore, overproducing the Na^+/H^+ antiport protein should provide increased salt tolerance in transgenic plants. Indeed, this hypothesis has been demonstrated in numerous studies in which researchers have generated transgenic rice, tomato, canola, corn, tobacco, and cotton, and in all cases the transgenic plants displayed enhanced salt tolerance in comparison with the non-transformed control plants. In transgenic tomato, it was shown that the salt does not accumulate in the edible fruit but is localized in the vegetative tissues. Under saline conditions, plants also cope with water loss caused by osmotic stress. By concentrating the salt in the plants' large vacuoles, not only the toxicity of Na^+ is avoided, but water that is free of salt is driven into plant cells. As a result, plants are able to use water more efficiently and are better equipped to withstand salt stress. This approach works in proof-of-concept models and greenhouse conditions but often requires combined strategies such as the overproduction of certain osmolytes or anti-oxidative enzymes to engineer salt stress tolerance in crop plants that can be grown on salty lands.

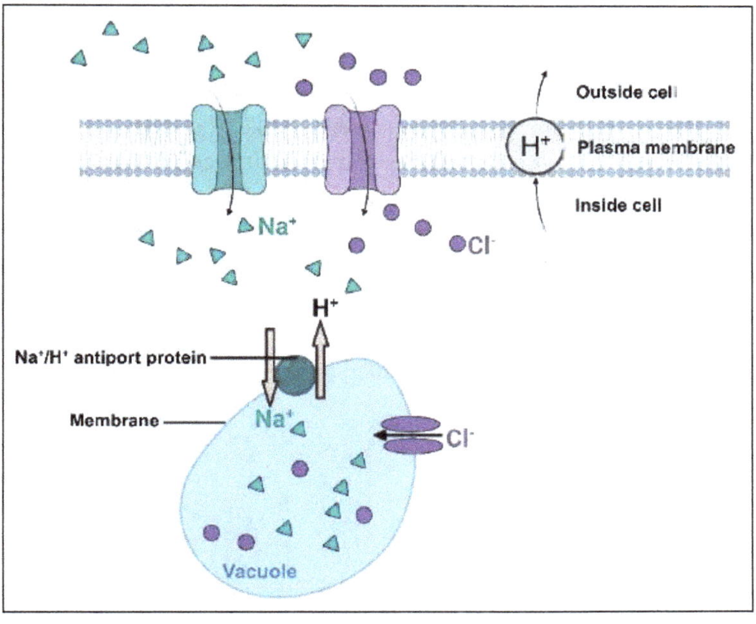

Fig. 6.20 An overview of ion transport. Na^+ enters the plant cell via channels and is sequestered into a vacuole by a Na^+/H^+ antiporter as a response to salt stress

Engineering trehalose accumulation

Trehalose is a naturally occurring highly stable molecule that is resistant to both acid and high temperature. Under desiccation (drying) or osmotic stress conditions, trehalose can form a gel-like matrix inside cells, thus decreasing the damage to cells from drought and salt. Unlike bacteria and yeast, in most higher plants, endogenous trehalose levels are low and insufficient to confer significant stress protection. However, biosynthetic enzymes for trehalose metabolism have been introduced into plants and this has often improved drought and salt stress tolerance. Generally, trehalose is synthesized by a two-step process: trehalose-6-phosphate is first formed from uridine diphosphoglucose and glucose-6-phosphate mediated by trehalose phosphate synthase (TPS). Trehalose-6-phosphate is then converted to trehalose by trehalose-6-phosphate phosphatase (TPP) (Fig. 6.21a). The *Escherichia coli* genes encoding the two enzymes that catalyze the two steps in the biosynthesis of trehalose were expressed together as a fusion protein in rice plants, thus simplifying the transformation process. Fig. 6.21b shows the binary constructs that were used for transforming the rice plants. In one T-DNA construct, the fusion protein gene is placed under the control of an abscisic acid-inducible promoter (where abscisic acid is the plant hormone that plays a central coordinating role in plant responses to both drought and salt stress) and is expressed in the cytosol. In the other construct, the fusion protein is under the transcriptional control of the *rbcS* gene (ribulose biphosphate carboxylase derived from the small subunit of Rubisco enzyme) and is expressed in the plant chloroplast. Under salt stress conditions, transgenic rice plants overexpressing either fusion gene construct demonstrated increased salt tolerance with 3-10 times higher trehalose content than the non-transformed plants. In addition, transgenic plants exhibited more prolonged photosynthetic activity than the non-transformed plants in the presence of salt, likely through maintenance of membrane stability due to the increased amount of trehalose. Thus, effective improvement in salt tolerance of crops is achievable by manipulating the trehalose biosynthesis pathway.

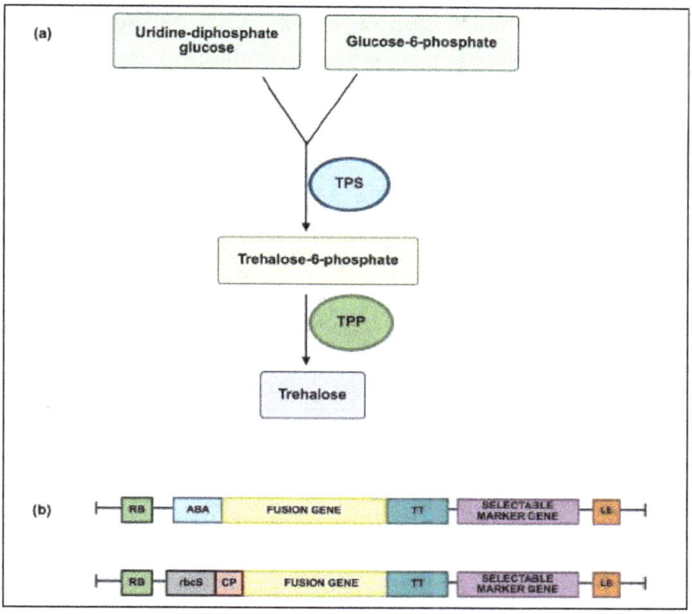

Fig. 6.21 Schematic representation of (a) trehalose biosynthesis pathway and (b) binary plasmids containing the trehalose biosynthetic fusion gene under the transcriptional control of an abscisic acid (ABA)-inducible promoter (top construct), or the promoter from the small subunit of ribulose biphosphate carboxylase (*rbcS*) (bottom construct). In the bottom panel, the transit peptide (CP) facilitates the targeted expression of the fusion protein in the chloroplast. Abbreviations: TPS-trehalose phosphate synthase, TPP-trehalose phosphate phosphatase, RB-right border, LB-left border

Drought-induced senescence

Delaying the onset of drought-induced senescence in plants contributes to enhanced drought tolerance by maintaining photosynthetic capacity during growth under water deficit conditions. In tobacco, leaf senescence could be delayed by overexpressing isopentyl transferase (IPT), an enzyme that catalyzes cytokinin synthesis. The IPT gene was put under the control of a senescence-associated receptor kinase (SARK) promoter, a maturation and a stress inducible promoter, and expressed in tobacco plants. The transgenic plants accumulated cytokinin specifically in senescing leaves where they effectively delayed the onset of drought-induced senescence through the suppression of senescence associated gene expression and maintaining photosynthesis during drought stress. Moreover, the transgenic plants continued to grow well even when watered with only 30% of the amount of water used under normal conditions. This may be a useful strategy to develop crops that can grow with substantially less water than is usually used, considerably minimizing drought-related losses.

Abiotic stress tolerance (including salt and drought) is a complex trait involving many genes and physiological processes. Scientists use microarray technology to identify genes that are upregulated when plants encounter such stresses, including genes involved in signaling, defense and adaptive responses and tolerance. For example, this approach is being used to identify genes when a plant is exposed to drought stress (Fig. 6.22).

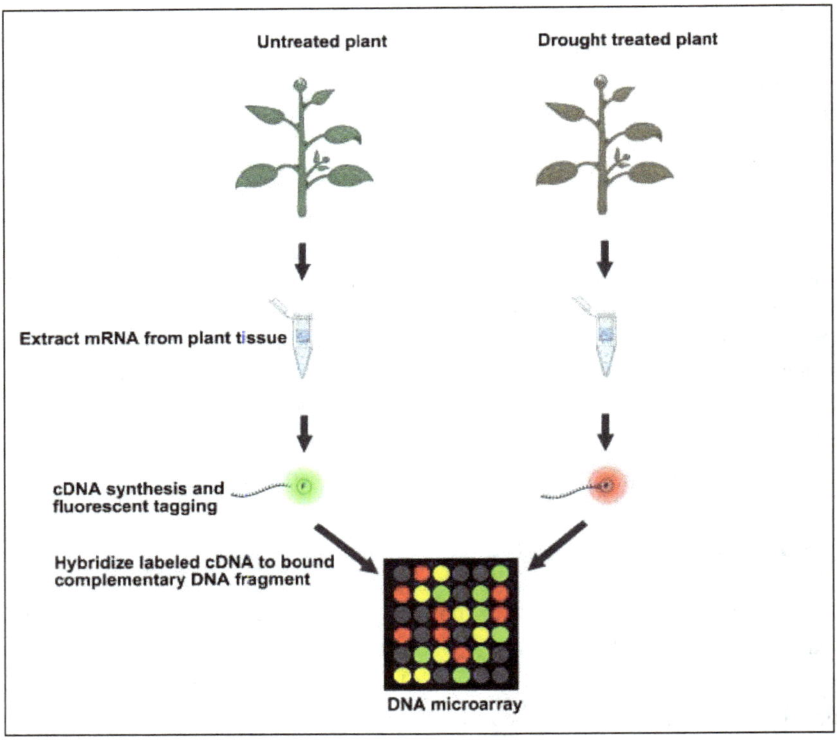

Fig. 6.22 Schematic overview of a DNA microarray experiment to study the plant gene expression pattern under drought stress conditions. Tissues are isolated from untreated, and drought treated plants; the mRNA is extracted from those tissues; using the mRNA as a template, cDNA is synthesized and then labelled with a fluorescent dye; the labelled cDNAs are hybridized to complementary DNA fragments bound to a microscopic slide; finally, the dye fluoresces, and the hybridizing sequences are identified by their fluorescent signals

6.6 Manipulating quality

Prolonging shelf life

Much of the food produce is lost as it travels from farmer to consumer. A delay in ripening and a longer storage life would be advantageous because it helps to preserve the nutritional qualities of produce for longer time periods and improve post-harvest quality. One way to achieve this is through genetic modification of genes involved in the ripening process. Experiments with tomato show that many changes occur during ripening. These changes affect all cell components and include a rise in respiration, softening of the cell walls, and synthesis of chemicals responsible for flavor and aroma. In particular, expression of the gene encoding the major cell wall softening enzyme, the enzyme polygalacturonase (PG), increased greatly during ripening. Indeed, the silencing of the gene encoding PG using an antisense approach modified the tendency of transgenic tomatoes to soften during storage, significantly reducing the PG mRNA and enzyme activity, subsequently delaying the ripening process (Fig. 6.23), allowing the tomatoes to slowly ripen on the vine (providing better flavor and color) instead of being harvested while they were green. This transgenic tomato variety known as the Flavr Savr ™ tomato maintains its firmness over an extended time period, making it less likely to bruise and thereby extending its shelf life. In 1994, this was the first genetically modified food product that was granted safety clearance by the U.S FDA and sold in U.S supermarkets. However, the Flavr Savr™ tomato was withdrawn from the market a few years later because it was not commercially viable.

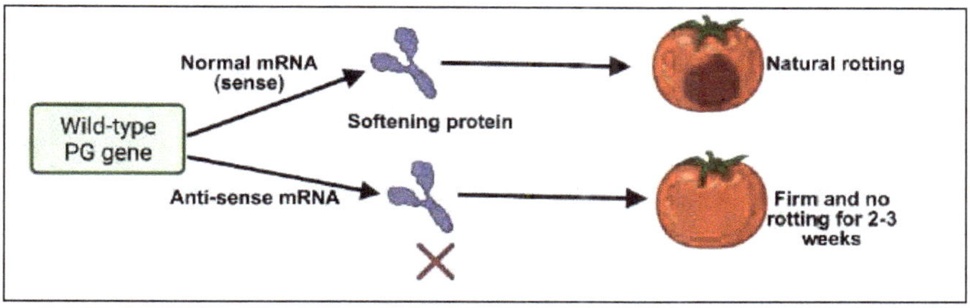

Fig. 6.23 Schematic representation of tomato fruit rotting due to the activity of the endogenous gene encoding the enzyme polygalacturonase (PG) (top panel). By using an antisense approach, the expression of the undesirable PG gene has been reduced, resulting in red, but firm fruit (bottom panel)

The plant hormone ethylene regulates fruit ripening and flower senescence. The biosynthesis of ethylene has been blocked using different strategies to delay fruit ripening and flower wilting. For instance, a strategy to generate transgenic plants having antisense ACO (aminocyclopropane-1-carboxylic oxidase) or ACC (aminocyclopropane-1-carboxylic) synthase gene has been implemented. The resultant transgenic plants synthesized much lower levels of ethylene than the non-transformed plants, and the fruit of the transgenic plants had a significantly longer shelf life. An approach has been used to delay ripening in tomato by expressing an ACC deaminase (*acdS)* gene isolated from a soil bacterium. This procedure requires screening soil bacteria for the ability to utilize ACC as a sole source of nitrogen, isolation of the gene for the enzyme ACC deaminase and the ability of the transformed *E. coli* strains expressing this gene to grow on selective media containing ACC as the sole nitrogen source. The gene is subcloned, put under the control of the constitutive 35S plant promoter, and expressed in tomato plants. The results demonstrated that the transgenic tomato plants produced less ethylene and as a consequence fruit ripening was delayed. An overview of the screening method for the selection of ACC degrading bacteria is shown in Fig. 6.24. A large number of plants have been engineered to enhance storage life including tomato, broccoli, apple, banana, melon and in flowers including carnations, petunia, rose, orchids, and tulips.

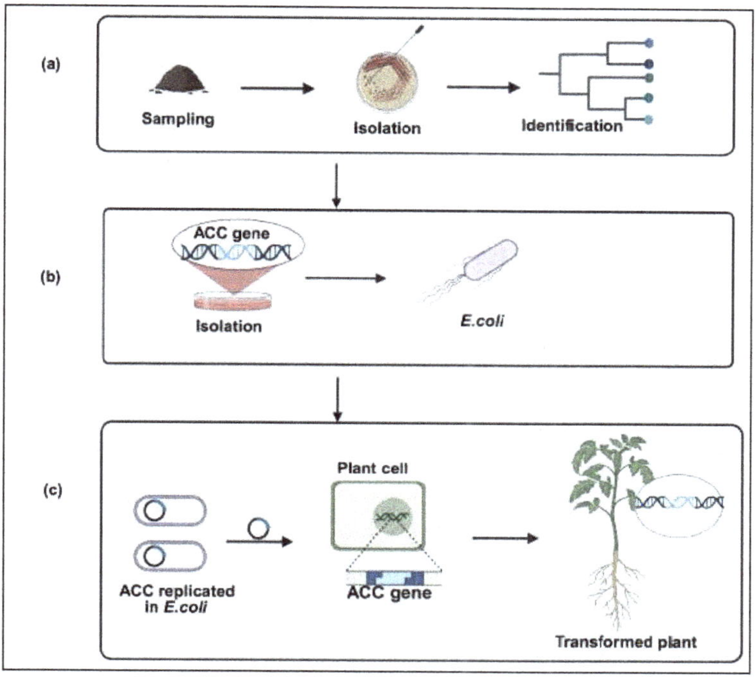

Fig. 6.24 General scheme for the selection of ACC-degrading bacteria and expression of ACC deaminase in plants (a) soil collection, screening, enrichment of ACC-degrading bacteria and identification (b) isolation of the ACC deaminase gene and selection of *E. coli* strain (c) subcloning and plant transformation

The shedding of petals is a natural phenomenon in flowering plants that can be accelerated by ethylene. In potted ornamental plants, controlling petal abscission is a desirable trait because it directly affects the shelf life and aesthetic appeal of the product. An attempt to extend flower life in *Torenia fournieri* (wish bone flower) by downregulation of the *aco* (ACC oxidase) gene resulted in transgenic plants which senesced later than normal, with a flower longevity of seven days compared to two days in the non-transformed plants. Also, the transgenic torenia plants with extended flower life produced more flowers simultaneously per stem than did the non-transformed plants, thereby improving their market potential.

It is possible to extend the shelf life of produce by using the CRISPR/Cas9 system as shown in Table 6.5. Targeted alterations of specific genes by the CRISPR/Cas9 method led to mutants with prolonged shelf life. In many cases, the mutation introduced by the CRISPR/Cas9 system is quite stable. Because crops developed by introducing synthetic sequences transcribing antisense RNA are categorized as genetically modified crops, they are tightly regulated and require costly safety evaluation. On the other hand, techniques such as CRISPR/Cas9, which do not introduce any foreign DNA into the plant, can mitigate some of the negative regulatory concerns associated with the public's perception of GM food crops.

Table 6.5 Utilization of CRISPR/Cas9 for prolonging shelf life of vegetables, fruits, and ornamentals

Plant	Target gene
Tomato	ALC, PL, PG2a, TBG4, SlDML2, RIN, MADS-RIN, TBG4
Potato	StPPO2
Eggplant	SmelPPO1-10
Sweet potato	GBSSI, SBII
Kale	MaACO1
Banana	MaACO, MADS1 and MADS2
Melon	CmACO1
Strawberry	PG1
Apple	PG1
Grapefruit	PPO
Petunia	PhACO, PhACO1, EPHI

Prevention of browning in fruits and vegetables

Browning of fruits and vegetables arises during post-harvest procedures and storage, while some pre-cut fruit and vegetables can turn brown within minutes of being cut and exposed to the air. Polyphenol oxidases (PPOs) are mainly responsible for the browning, and they are naturally present in fruits and vegetables. When a fruit or vegetable is cut or bruised, in the presence of oxygen, PPOs catalyze the oxidation of phenols to quinones (Fig. 6.25). This undesirable trait has become a challenge in the food industry as the damage caused due to browning can affect taste, texture as well as the appearance of the produce. Additionally, browning often leads to food being unnecessarily thrown away because people mistakenly believe browned food is inedible. To solve this problem, the food industry adds browning inhibitors which can negatively affect color, taste, flavor, and nutritional value. Moreover, some of these chemical additives may pose a risk to human health.

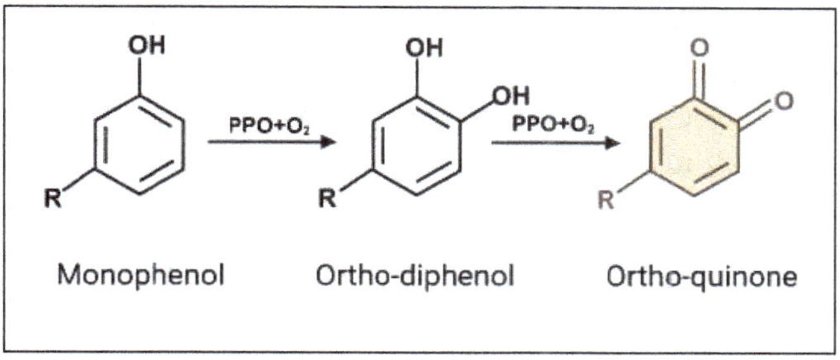

Fig. 6.25 Schematic representation of an enzymatic browning reaction. Polyphenol oxidase (PPO) oxidizes phenolic substrates in the presence of oxygen resulting in the formation of a brown *o*-quinone product. R in the figure represents any functional group

Scientists have focused on engineering crops to downregulate PPO genes to reduce enzymatic browning. One way this can be done is by the production of antisense transcripts. Inverting a target DNA sequence, relative to its promoter, produces mRNA transcripts which are complimentary to the target gene transcript. Since the transgene is driven by the promoter sequences of its native counterpart gene, expression should be limited to that location where it can reduce the amount of targeted gene product. This has been the key strategy in developing commercial non-browning cultivars of potato and apple. In one study, a PPO cDNA fragment was cloned into a vector either in sense or antisense orientation (Fig. 6.26). Expression was driven under a promoter, significantly stronger (about 100 times) in potato tubers than in leaves. When assessing the resistance to browning, transgenic plants (with the abovementioned anti-sense construct) that were subjected to deliberate bruising, were significantly more resistant to black spot (discoloration by PPO), resulting from fewer PPO proteins being produced. The sense constructs produced more of the PPO proteins and displayed extensive black spotting than the non-transformed control plants. This study was key to the development of commercial varieties of genetically modified potato and apple with non-browning traits.

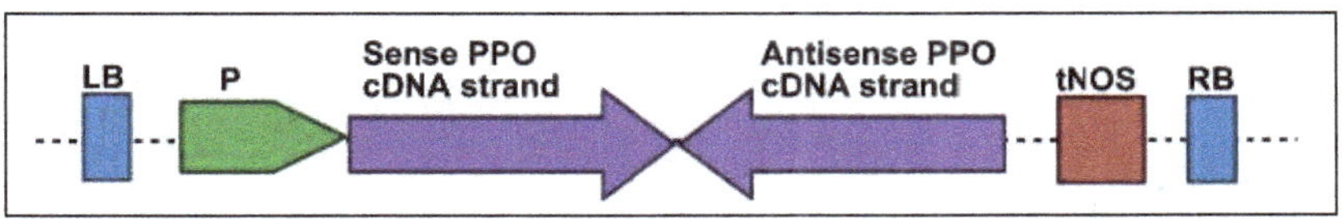

Fig. 6.26 Sense and antisense polyphenol oxidase (PPO) gene constructs. Promoters are in sense and antisense orientation respectively, each under separate control (35S and tissue-specific) of a promoter (P) with a nopaline synthase transcription terminator region (tNOS); RB and LB, right and left borders of the T-DNA, respectively

Understanding PPO biosynthesis in greater depth has allowed development of other crops that suffer enzymatic browning such as eggplant and mushrooms. Using the CRISPR/Cas9 system, scientists engineered an edible mushroom (*Agaricus bisporus)* to resist browning. Six genes responsible for PPO activity were knocked out in the mushroom, decreasing the enzyme activity by 30%. In a similar study, three PPO genes in eggplant have been targeted to reduce enzyme activity, and the resultant eggplants exhibited reduced browning in the flesh. However, in polyploid crops such as potato, CRIPSPR/Cas9 editing can be technically challenging as copies of each gene must be edited to ensure all alleles are modified. If a conserved region is not found, a different sgRNA will need to be designed. Nevertheless, mutation of the StPPO2 gene in potato targeted by a ribonucleoprotein complex consisting of two sgRNAs and Cas9 nuclease was performed. The total PPO activity in the modified potato tubers was reduced by 69% resulting in a 73% reduction of enzymatic browning.

Color modification

Pink fleshed pineapple

In 2016, a company in the United States developed a genetically modified pineapple (known as Pinkglow™) to increase lycopene levels in the fruit's flesh to produce a pink color. Lycopene (a carotenoid and powerful antioxidant) is naturally found in pineapples, and it is the pigment that makes tomatoes red and watermelons pink. Lycopene is synthesized via a carotenoid metabolic pathway. During fruit development, lycopene is converted to β-carotene through the action of the LCY-E (lycopene ε-cyclase) and LCY-B (lycopene β-cyclase) genes. The developers of this technology overexpressed the PSY gene (a rate limiting step in carotenoid biosynthesis) from tangerine and suppressed the cyclase genes using RNAi technology in order to increase accumulation of lycopene levels in the flesh of pineapple fruit.

Blue carnation

Blue flowers exhibit color due to the presence of delphinidin, a type of anthocyanin. Carnations (*Dianthus* spp.) are economically valuable flowers that do not naturally produce delphinidin. Carnations have been genetically engineered to make this pigment and create blue flowers, offering something novel and unique for consumers. The key enzyme in the biosynthesis of delphinidin is flavonoid 3',5'-hydroxylase (F3'5'H). Heterologous expression of a exogenous petunia F3'5'H gene has shown to direct production of a blue hue in petunia flowers. Similarly, a shift in the flower color was also observed when a F3'5'H gene isolated from Canterbury bells (*Campanula medium*) was introduced in petunia. The transgenic flowers were bluer with a greater accumulation of delphinidin, reaching up to 99% of total anthocyanins, suggesting a more efficient enzyme activity of the *Campanula* F3'5'H. Thus, it is possible to genetically manipulate the chemical composition of flowers to produce novel ornamental plants.

Purple tomato

Purple tomatoes are one of the many color varieties that are available on the market, which other than their unique color, are preferred for their health benefits. Because of the accumulation of anthocyanin (an antioxidant) in their skin, they appear purple, rather than red. To produce purple tomatoes, researchers transformed the wild-type tomato with the Delila (Del) and Rosea (Ros1) genes from snapdragon (*Antirrhinum majus*), an ornamental plant. A binary vector containing the Del and Ros1 cDNAs, each put under the control of the fruit specific promoter E_8, was used to transform tomato leaf discs via *Agrobacterium*-mediated transformation (Fig. 6.27). The amount of anthocyanin accumulated was increased up to 2.8-fold in transgenic fruits compared to non-transformed fruit, clearly demonstrating the strong effect. Anthocyanin expression became progressively more intense as the fruit ripened with pigmentation, first visible in the vascular tissue and then extending to peel, pericarp, and inner flesh.

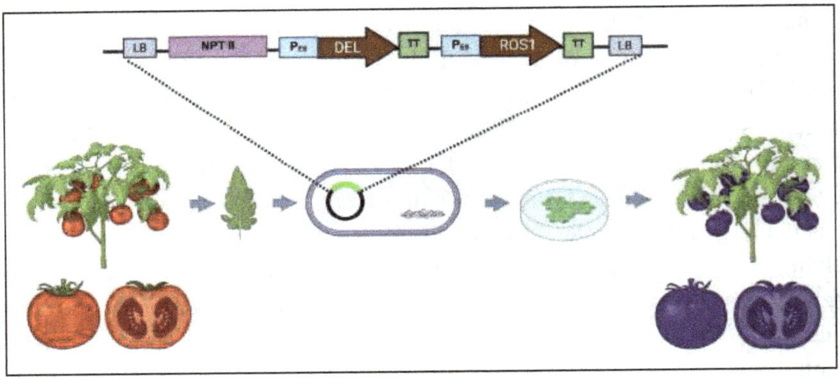

Fig. 6.27 Illustration of the Ti plasmid vector used to transform tomato and produce the purple pigment in the transgenic fruit. Each of the two genes - Del and Ros1, is expressed under the control of promoter E_8 that causes fruit-specific expression. The T-DNA also includes the NPTII selectable marker, TT, transcription termination region; LB, left border; RB, right border. The tomato plant is transformed via *Agrobacterium* using the leaf disc method (see Chapter 5)

Improvement of nutritional properties

Proteins

Humans and animals are unable to synthesize nine out of twenty essential amino acids, which must therefore be acquired via dietary intake. Seeds grown for food (e.g., cereals and legumes) contain large amount of protein and are a good source for the nutrition of humans and livestock. However, they often contain a limited number of certain amino acids. Altering the amino acid profile in plants by conventional breeding can be difficult and slow. An alternative strategy would be to transform crops with genes encoding enzymes that would enhance the amino acid composition of proteins in seeds.

Lysine, an essential amino acid, is important for both human and animal health. To increase the lysine levels in seeds, manipulating the lysine biosynthesis pathway seems a promising approach. In plants, lysine is synthesized from aspartic acid through a two-step enzymatic process (Fig. 6.28). In the first step, the enzyme aspartate kinase (AK) phosphorylates aspartic acid to produce β-aspartyl phosphate which is then converted to aspartic β-semialdehyde. Next, the dihydrodipicolinic acid synthase (DHDPS) catalyzes the conversion of aspartic β-semialdehyde to 2,3-dihydrodipicolinic acid. This step is considered a crucial step for lysine biosynthesis. Both AK and DHDPS are extremely sensitive to feedback inhibition by lysine and represent the major limiting step for the synthesis of this amino acid. Thus, to overproduce lysine, scientists have used various approaches to suppress the feedback inhibition of these two enzymes. For example, bacterial feedback-insensitive AK and DHDPS were tested for the ability to improve lysine content. In this work, scientists cloned genes for DHDPS and AK from *Corynebacterium* and *Escherichia coli,* respectively; fusing each of these genes to a chloroplast transit peptide to direct the bacterial enzymes to seed plastids; placing each gene under the control of a seed-specific promoter; and then introducing the chimeric genes on a Ti plasmid binary vector into canola and soybean plants. As a result, the lysine content was increased by more than 100-fold in transgenic canola and soybean seeds. This strategy has also been successfully employed in rice, maize, and barley.

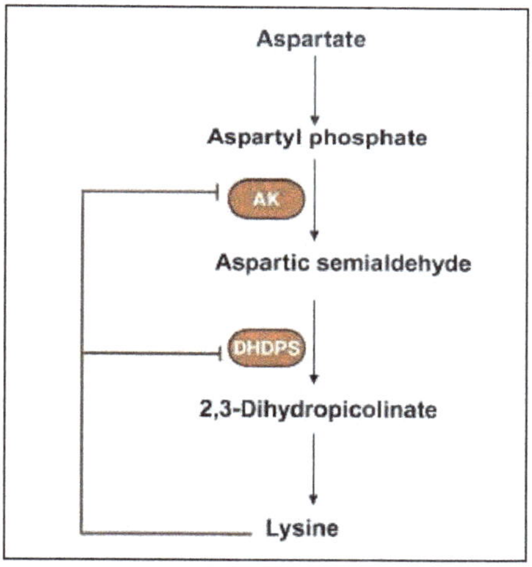

Fig. 6.28 Lysine biosynthesis via aspartate metabolism pathway in plants. Brown lines indicate product feedback inhibition

Lupine, a grain legume, has very low levels of the sulfur-containing amino acids methionine and cysteine. To improve its nutritive value as an animal feed, lupines were engineered to express a sunflower seed albumin, which is both stable in the rumen and exceptionally high in methionine. The resulting transgenic plants had elevated methionine levels, and when fed to rats this enriched lupine promoted increased weight gain compared to the non-transformed lupine seeds, demonstrating improved protein quality. Moreover, diets based on methionine resist microbial breakdown in the rumen gut and can improve feed efficiency while matching the amino acid needs of ruminants.

Wheat grains are rich in gluten, a storage protein, that influences the elastic properties of wheat dough which is important for bread making. Gluten is not a single polypeptide but a complex mixture of different classes of proteins, mainly prolamins. Gliadins and glutenins make up the majority of prolamins in wheat grains and are important because they are associated with desirable bread making traits. The amount and composition of these proteins has been altered via genetic engineering resulting in wheat plants with novel seed protein compositions. Higher molecular weight glutenin polymers are desirable because they help the dough to rise and maintain its shape, often giving the product a chewy texture. Accordingly, high molecular weight glutenin genes have been transferred to a number of wheat varieties, with improvement in the functional properties of wheat flour, resulting in the aforementioned qualities. However, gluten proteins, especially gliadins, are also associated with celiac disease (CD), a gastrointestinal disorder characterized by lesions in the small intestine, causing diarrhea, fatigue, and malabsorption of food. Prolonged gluten consumption can cause other symptoms including osteoporosis, and neurological symptoms including headaches. This disease is prevalent in ~1-2%, of the global population, and the only treatment until now is a strict, life-long gluten-free diet. On the other hand, through genetic engineering, it is possible to develop wheat varieties that may be tolerated by many CD-patients. However, this manipulation is challenging due to the genetic complexity of gliadins. Despite this, a group of scientists has engineered bread wheat using RNAi technology with hairpin vectors to simultaneously silence multiple gliadin genes. These vectors consisted of a promoter and terminator region flanking an inverse repeat sequence of the target gene, with a spacer region between the repeats such that when transcribed in the plant cell, the products form a hairpin by intra-strand complementary base pairing (Fig. 6.29). The targeted gene included a 361bp DNA segment that is highly conserved among the wheat gliadin family. Plants transformed with these vectors showed a strong decrease in the gliadin protein content. Therefore, this work could be the basis for the development of products suitable for individuals with gluten-intolerance. As a step forward towards the development of non-transgenic wheat varieties with decreased levels of gliadins, another group of scientists engineered the wheat genome with a CRISPR/Cas9 tool. This approach resulted in a significant reduction in α-gliadin proteins in the wheat kernels, with a reduction in immunoreactivity of 85%.

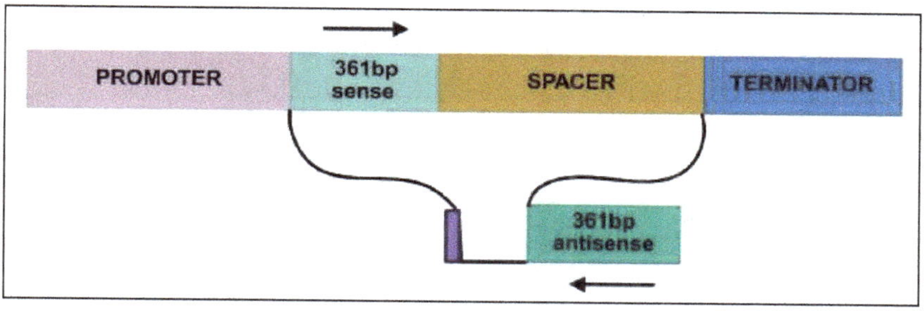

Fig. 6.29 Schematic diagram of the hair pin construct which includes a promoter region, the sense strand of the 361bp target DNA, a spacer, the antisense strand of the 361bp target DNA, and a transcription terminator region. The hairpin structure and arrows indicating the sense and antisense orientation are also shown

Oils

Plants accumulate oils during seed development. Plant oils are primarily composed of triglycerols (TAG) (Fig. 6.30), with fatty acids (FA) and glycerol extracted during downstream processing.

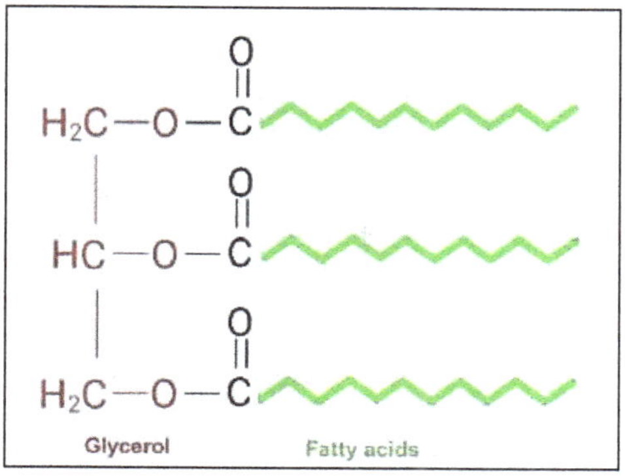

Fig. 6.30 Structure of triglycerol

The FAs may vary in the number of carbons in the chain, number and position of double or triple bonds, and the presence of epoxy, hydroxy, and other functional groups. In saturated fatty acids (SFA), the C-C atoms are connected together with a single bond whereas in unsaturated fatty acids (UFA) they are linked together with a single or two or more double bonds (Fig. 6.31).

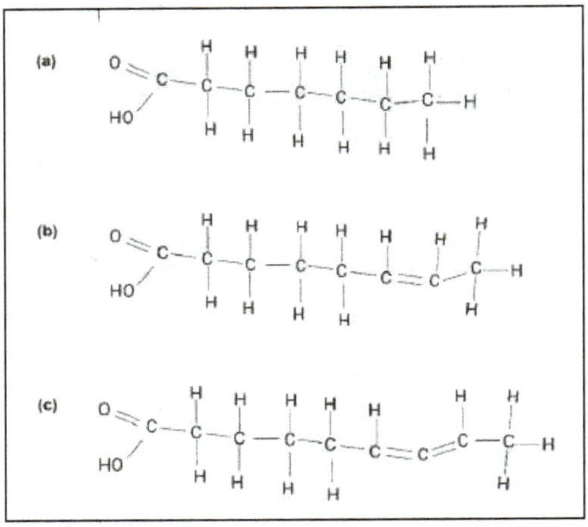

Fig. 6.31 Structures of FAs (a) saturated fatty acid (SFA) (b) monounsaturated fatty acid (MFA)with one C=C bond (c) polyunsaturated fatty acid (PFA) with two C=C bonds

Oilseed crops serve as food sources to humans and animals. Moreover, they have emerged as sustainable alternatives in many industrial applications. The major oil products are obtained from soybean *(Glycine max),* canola (*Brassica napus*), sunflower seed *(Helianthus annuus)*, peanut *(Arachis hypogaea)* and cottonseed *(Gossypium hirsutum)*. These oils contain various proportions of common fatty acids including palmitic, stearic, oleic, linoleic, and α-linolenic acids (Table 6.6).

Table 6.6 Range of oil content and fatty acid profile of selected oilseeds (a) saturated fatty acid (b) monosaturated and polysaturated fatty acid

Oilseed	Oil content (%)	SFA (%)	MUFA (%) Oleic ($\Delta 9C_{18:1}$)	PUFA (%) Linoleic ($\Delta 9,12C_{18:2}$)	α-linolenic $\Delta 9,12,15C_{18:3}$)
Soybean	18-28	15	23	56	8
Canola	40-65	7	61	21	10
Sunflower	40-50	12	16	63	-
Peanut	45-62	19	48	33	-
Cottonseed	15-25	27	20	54	-
Palm	45-55	51	39	10	-
Coconut	60-65	90	6	2	-
Olive	15-30	15	75	10	1

Note: The number after the C denotes the number of carbon atoms; the number after the colon is the number of C=C bonds; a number after the Δ indicates the position of the first carbon atom in C=C; with the carboxyl group numbered as one

The fatty acid composition (chain lengths, degree of saturation, and other structural features) determine the functional, sensory, and nutritional value of edible oils. Identification and functional characterization of genes involved in pathways of oil synthesis has led to the development of a variety of plant seed oils tailored for specific applications. For example, the fatty acid desaturase 2 (FAD2) enzyme normally converts oleic acid into linoleic acid. Inhibition of FAD2 activity should favor oleic rather than linoleic acid production. Indeed, suppression of FAD2 activity in transgenic soybean increased the oleic acid content, with up to 86% in soybean. This modification improves the shelf-life and stability of oils in high temperature cooking applications. Similarly, high stearic acid producing crops are of significant interest because stearic acid rich oils are useful in the production of confectionary goods and margarine. Stearoyl-ACP desaturase (SAD) facilitates the desaturation of stearic acid to oleic acid by introducing a double bond. A transgenic variety of canola containing an antisense version of a *Brassica* SAD gene, which inhibits the expression of the normal canola gene, led to an accumulation of stearic acid in seeds by blocking the conversion of stearic acid to oleic acid.

Very-long-chain polyunsaturated fatty acids (VLC-PUFAs) including omega-3 and omega-6, are nutritionally important fatty acids with many health benefits. They have been implicated in lowering blood pressure, reducing triglycerides, reducing the risk of heart disease, and improving other human and animal health conditions. Until now, these fatty acids have been derived from fish and marine microorganisms. However, it has been difficult to obtain a sufficient level of these fatty acids from these sources, due to declines in fish populations along with other problems associated with fish farming. VLC-PUFAs are produced from linoleic and α-linolenic acid (C-18 precursors of the omega-3 and omega-6 fatty acids). Since plants can synthesize these precursors, it may be possible to modify the synthesis pathway by introduction of exogenous genes from distant species into plants, to convert linoleic and α-linolenic acid into long chain omega-3 and omega-6 fatty acids. In one study, three genes from different species: a Δ9-specific elongase from the marine alga *Isochrysis galbana,* a Δ8-specific desaturase from the protist *Euglena gracilis*, and a Δ5-specific desaturase from the fungus *Mortierella alpina*, were expressed constitutively in *Arabidopsis thaliana* plants. The engineered plants that were generated by three rounds of transformation contained higher than normal levels of arachidonic acid (7%) (an omega-6 fatty acid) and eicosapentaenoic acid (EPA) (3%), an omega-3 fatty acid, in their vegetative tissue. Despite the storage oil having been altered, the expression of the genes did not affect the growth and development of the plants. Further progress has been made to reconstitute the complete VLC-PUFA pathway in transgenic plants as more enzymes are being identified. Transgenic tobacco leaves with total fatty acids containing up to 1.4% EPA were produced by expressing Δ6-desaturases from *Osterococcus tauri* (microalga), *Phaeodactylum tricornutum*, (diatom), *Rhizopus stolonifera* (fungus), Δ^5-desaturases from *P. tricornutum* and *Thraustochytrium* sp. (single-celled saprotrophic eukaryote) and adding an acyl-CoA:lysophosphatidylcholine acyltransferase (LPCAT) from *P. tricornutum* to maximize the accumulation of omega-3 fatty acid. Further efforts should be directed towards seed specific expression of these enzyme combinations in an oilseed crop such as canola or soybean.

Vitamins

Humans need vitamins and minerals (micronutrients) for mental and physical development. These are normally provided in a varied and nutritionally balanced diet. People who rely on a single crop may suffer from micronutrient deficiencies as many plants lack multiple nutrients. An important objective of a plant biotechnologist is, therefore, to biofortify crops such as rice, wheat, corn, banana, and potato so that they more closely meet our nutritional requirements. Limited success has been achieved with cereals by molecular breeding methods. Biofortification through genetic engineering includes (a) insertion of synthetic genes that encode proteins of nutritionally desirable composition synthesized *de novo* (b) improved uptake, transport, and storage mechanisms for nutrients acquired from the environment. A few most common micronutrient deficiencies are discussed below.

Vitamin A

Vitamin A deficiency is a critical world health problem, particularly affecting children. While illnesses from vitamin A deficiency often cause vision impairment (total blindness or night blindness), the increased risk of severe infection from a weakened immunity can also lead to mortality among children. To address this issue, researchers have developed several biofortification strategies targeting staple crops. Humans can synthesize vitamin A from the precursor molecule β-carotene (also known as provitamin A). Because approximately half of the world's population relies on rice as a staple diet, there has been a significant interest in engineering rice to produce provitamin A. To facilitate β-carotene synthesis in rice, scientists introduced the daffodil phytoene synthase (*psy*) gene, the phytoene desaturase (*crt*) gene from the bacterium *Pantoea ananatis* (formerly known as *Erwinia uredovera*), and the lycopene β-cyclase (*lcy*) gene from daffodil, which reconstituted the entire β-carotene biosynthetic pathway (Fig. 6.32) into the endosperm of rice. The *crt* gene was expressed under the control of the CaMV 35S promoter, while the *lcy* and *psy* genes were expressed under the control of the rice glutenin promoter. Each gene was fused to a transit peptide so that the proteins they encoded would be transported into the plastid. This engineered rice variety accumulated β-carotene, resulting in a golden color, which has been called "Golden Rice" (GR). This grain produced 1.6 μg of β-carotene per gram of rice. However, this line of rice, now called GR1, was criticized because of the large amounts of rice that would be needed to be consumed by an individual to meet the minimum daily requirement of vitamin A in humans. More recently, an advanced version of Golden Rice, GR2, in which the daffodil phytoene synthase gene was replaced with a similar gene from corn that encodes an enzyme with a higher level of activity, resulting in a 23-fold improvement in the β-carotene content. Based on these studies, other crops including golden potatoes, golden bananas, and golden cassava have been developed. In one biofortified golden potato cultivar "Desiree", the tubers contained 47 μg of β-carotene per gram of dry weight, sufficient to provide 50% of the recommended individual dietary intake. Using the CRISPR/Cas9 system to improve β-carotene levels in the "Grand Naine" Cavendish banana cultivar by editing the lycopene ε-cyclase gene, resulted in a 6-fold increase in β-carotene content within the fruit pulp compared to unedited plants.

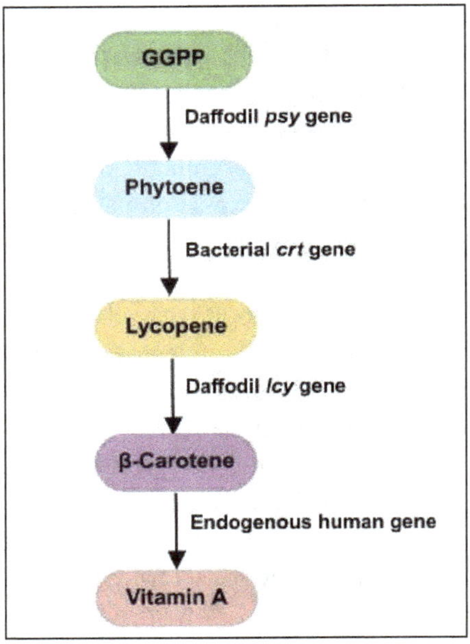

Fig. 6.32 Overview of the biosynthesis of β-carotene in rice and vitamin A in humans. GGPP, geranylgeranyl pyrophosphate, is converted by phytoene synthase into phytoene, which is then processed through several steps into β-carotene

Iron

More than 2 billion people worldwide suffer from iron deficiency. Given its widespread consumption, rice is a suitable target for iron biofortification. An early approach to engineer rice plants that produce enhanced levels of iron involved expressing the soybean ferritin (an intracellular protein that stores iron) gene using the rice endosperm specific promoter. The binary vector construct used to transform rice to increase iron content is shown in Fig. 6.33. Transformants showed 2.5-fold higher levels of iron accumulation in rice seeds per gram (dry weight). This endosperm-targeting expression of ferritin has recently been used with the *Indica* rice strain IR64 and *Japonica* rice strain Taipei 309 with the following increases in iron concentration compared to the non-transformed strain: a 3-fold increase in the transgenic *Indica* rice strain IR64 and a 2.2-fold increase in the *Japonica* rice strain Taipei 309. Several additional methods are now available to achieve higher iron accumulation in rice.

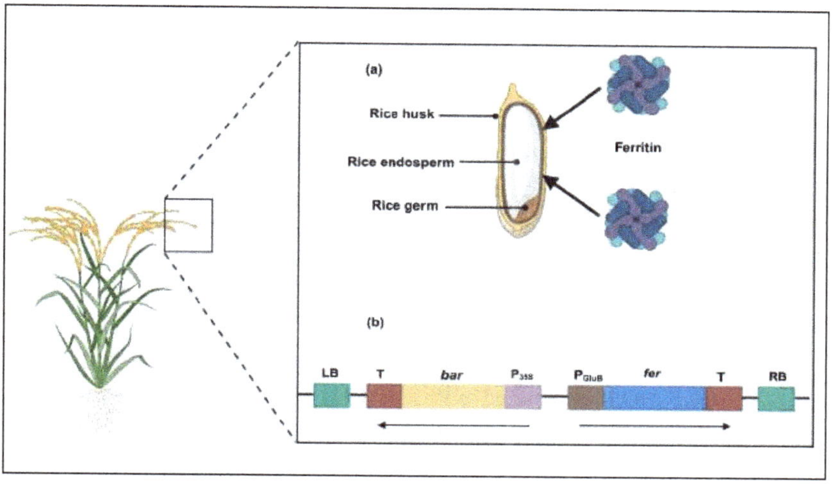

Fig. 6.33 A schematic representation showing a rice plant engineered to produce the soybean protein ferritin under the control of an endosperm-specific promoter. (a) a cross-section of a rice grain (b) the binary construct used to transform rice. The arrows indicate the direction of transcription. LB, left border; T, termination region; *bar,* the bacterial phosphinothricin acetyltransferase gene used as a selectable marker; P_{35S}, 35S promoter from *Cauliflower mosaic virus*; P_{gluB}, promoter from the rice seed storage protein glutelin; *fer,* soybean ferritin-encoding cDNA; RB, right border

Manipulation of Starch

Starch is the main storage carbohydrate in plants. The starch content in tubers of potato and yam, cassava roots and sweet potato, and endosperm of rice and maize is 65-90% of the dry weight of the storage organs. Starch is widely used in food and industrial applications. It is composed of amylose (20-30%), a linear α-1,4 glucan polymer and amylopectin (70-80%), a branched α-(1,4 and 1,6) glucan polymer. The ratio of amylose to amylopectin determines the physical and chemical properties of the starch granules. It is feasible to genetically modify plants for altered starch content and composition to produce starches with new properties for a range of specific applications. Starch biosynthesis takes place through a series of steps (Fig. 6.34a). The synthesis of amylose and amylopectin (Fig. 6.34b) is primarily regulated by three key enzymes in this pathway. The ADP-glucose phosphorylase initiates starch biosynthesis by catalyzing the transfer of ADP (from ATP) to glucose-1-phosphate to form ADP-glucose; the enzyme starch synthase catalyzes the elongation of the glucose chains by adding glucose from ADP-glucose to the non-reducing end of an existing α-1,4 linked glucan polymer. Branching enzyme converts amylose to amylopectin by breaking the α-1,4 linkages and synthesizing α-1,6 branch points.

Amylose-free (also called 'waxy') or high amylopectin starches can be generated using RNAi and gene editing technologies. This trait is desirable in the food processing industry because amylose-free starch paste is clear and stable, and as such, is the preferred form of starch for use as texture modifiers in many food products. Transgenic potato plants were generated using antisense technology by switching off the granule-bound starch synthase (GBSSI) gene which is involved in amylose synthesis. As a result, the starch in the tubers was free of amylose which correlated with the loss of the GBSS enzyme activity. Because of their lack of amylose, these starches have superior freeze-thaw stability which are advantageous as they reduce syneresis (a process where starch separates into a solid phase and a liquid phase) and maintain the gel integrity during freezing and thawing of frozen food. More recently, this gene was targeted for knockout using the CRISPR/Cas9 system in the Yukon Gold variety of potato, resulting in low amylose content tubers. The potato studies offer a model system to modify starches with novel qualities that can be used to develop new applications in the food industry.

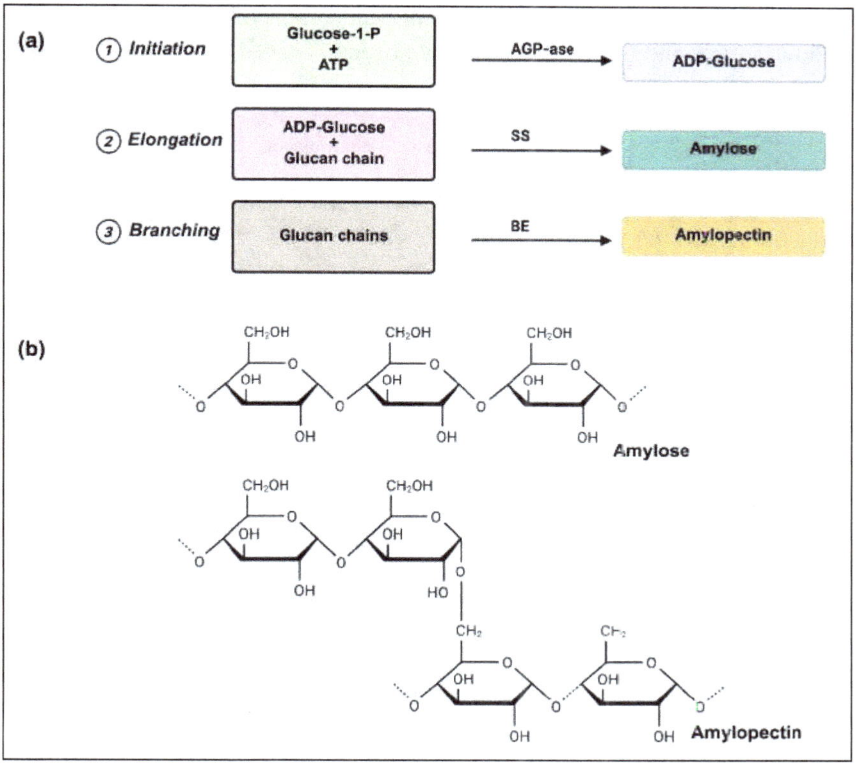

Fig. 6.34 Overview of (a) steps involved in biosynthesis of starch (b) structures of amylose and amylopectin. Abbreviations: AGP-ase, ADP-glucose pyrophosphorylase; SS, starch synthase; BE, branching enzyme

6.7 Plants as bioreactors

Advances in plant biotechnology have boosted commercial interest in using plants as an expression platform for the production of recombinant proteins (especially human pharmaceuticals). Plants are able to produce nearly any protein. Moreover, compared to mammalian or bacterial platforms, plants have the advantage of producing proteins that are both safe and biologically active, at a significantly lower cost. Since plant products have long been used in medicine, the technology is already available for harvesting and processing plant synthesized products on a large scale. In fact, the ability to produce large quantities of human pharmaceutical products in plants as bioreactors has allowed production of a number of therapeutic compounds, some of which are in the midst of clinical trials and a few of which have been approved for human use (Table 6.7).

Table 6.7 Plant synthesized pharmaceuticals

Product	**Disease**	**Plant**
Spike protein ACE_2FC, IBIO-201, IBIO-400, MT-2766	SARS-CoV-2	Tobacco
PRX_{102}	Fabry disease	Carrot
Elelyso (taliglucerase alfa)	Gaucher disease	Carrot
rhIGF-1 LR3	Growth factor	Rice
OsrhLF	Human lactoferrin	Rice
Zmapp	Ebola virus	Tobacco
HIV microbicide MB66	HIV	Tobacco
Norwalk virus capsid protein	Diarrhea	Potato
ISOkine	Growth factor	Barley
Heat-labile (Lt-B) toxin of *E. coli*	Diarrhea	Potato, corn
VEN BETA	Gastroenteritis	Rice
VEN-120	Inflammatory bowel disease	Rice
Optibumin	Loss of albumin	Rice
DPP4-Fc	Middle East Respiratory Syndrome (MERS)	Tobacco

Whole plants or specific plant parts can be used to produce recombinant proteins. Tobacco is often a preferred plant for the synthesis of foreign proteins because of its large leaf size; seed crops are suitable because the pharmaceuticals can be synthesized in kernels, which can readily be processed to create a purified sample. Recombinant proteins can also be produced in edible plant tissues engineered for use as a food additive or in a capsular form or as an adjuvant in vaccines. There are various strategies to create transgenic plants that synthesize recombinant proteins including therapeutic, industrial or immunogenic purposes. Fig. 6.35 illustrates one such approach.

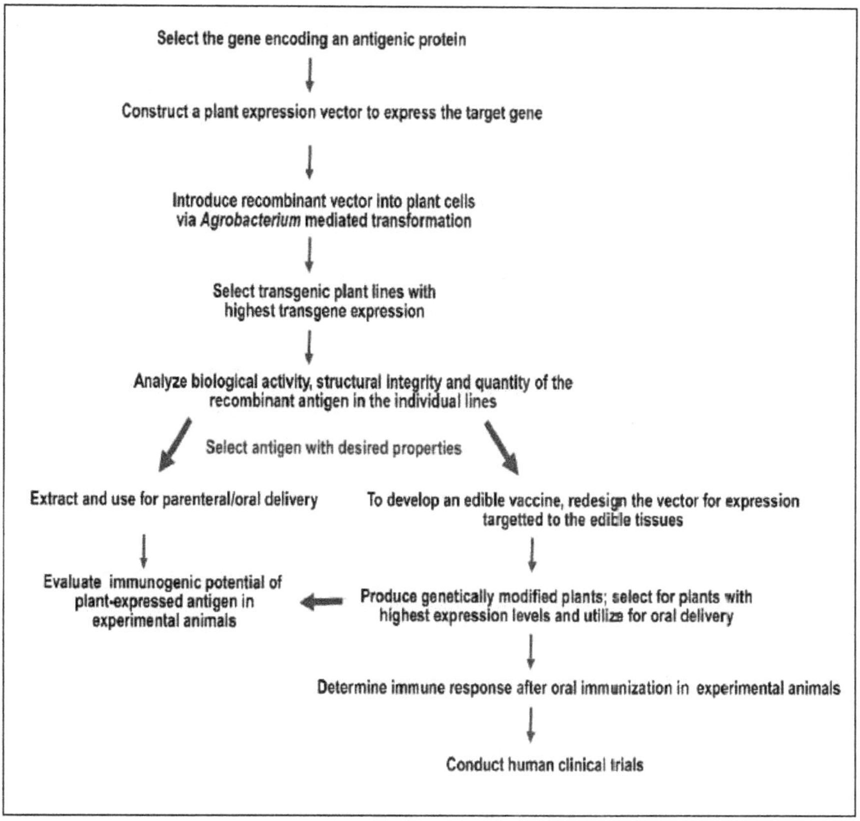

Fig. 6.35 Flow chart summarizing the general steps used in the production of immunogenic proteins in transgenic plants

Production of antibodies

A large number of different human monoclonal antibodies (mAbs) including immunoglobulins IgG, IgA, IgM, and single chain Fv fragments and Fab fragments have been produced in plants for protection against different pathogens (Table 6.8). The first plant-derived antibody (sometimes called a plantibody) was synthesized in tobacco plants and was directed against the bacterium *Streptococcus mutans,* that causes tooth decay. This antibody conferred protection against the tooth decay causing bacterium in human trials.

Table 6.8 Examples of potentially therapeutic and diagnostic antibodies and antibody fragments that have been produced in transgenic plants

Antibody	Plant host	Target disease
SIgA	Tobacco	Dental caries
IgG	Soybean	Herpes simplex virus
ScFv	Wheat	Cancer
ScFv	Rice	Cancer
IgG	Tobacco	Cancer
ScFv	Tobacco	B-cell lymphoma
IgG	Tobacco	Colon cancer
IgG	Corn	Respiratory virus
IgG	Tobacco	Rabies
IgG	Tobacco	Anthrax
IgG	Tobacco	Hepatitis
IgG	Tobacco	Botulism

Plant (leaves and seeds) can express genes encoding the protein subunits of mAbs and assemble them into functional complexes. An important property of plant cells is that the processing and assembly of foreign proteins are similar to those in animal cells. On the other hand, bacteria cannot perform post-translational modifications and often yield misfolded proteins. As in many production systems, a critical aspect is the recovery of the protein that is produced. Most bacteria are transformed with plasmids that may be lost during large-scale production. However, in the case of plant transformation, the foreign DNA is stably integrated into the plant genome. However, one problem associated with using a plant expression system is that the product yield is often very low (usually 1-10 μg/g of fresh biomass), which may not be sufficient to meet therapeutic requirements. Fortunately, many methods now exist to increase yields. One means of doing this is based on the use of a transient expression system (Fig. 6.36a), which is both rapid and expresses the target protein at a high level. One transient expression system involves co-infection of plant cells with two non-competing virus vectors, one from *Tobacco mosaic virus* (TMV) and the other from *Potato virus X* (PVX). Since the synthesis of monoclonal antibodies in plants requires the expression of two types of polypeptides (heavy and light chain), both vectors coreplicate and drive high-level expression of its respective chain, i.e., one vector expresses the antibody light chain and the other expresses the heavy chain (Fig. 6.36b). In this case, the total yield of the assembled IgG protein was found to be 0.5 mg/g of fresh leaf biomass. This technology has tremendous potential because it is much less expensive to produce antibodies from transgenic plants than from bacteria or mammalian cells which require costly units such as fermentation reactors or sterile mammalian culture vessels notwithstanding the fact that plants must be grown in controlled greenhouses, that can allow for their mass production and containment. It is noteworthy that both TMV and PVX have a wide host range thereby allowing for the use of multiple plant species as bioreactors. On the other hand, it is important to keep in mind that, just as with the production of mAbs in bacteria or animal cells, extraction of antibodies from green leaves requires purification. However, recombinant proteins might be produced in dry tissues such as seeds, as they are less likely to contain unwanted impurities. Additionally, seeds have high protein stability, allowing storage for extended periods of time under ambient conditions without loss of activity.

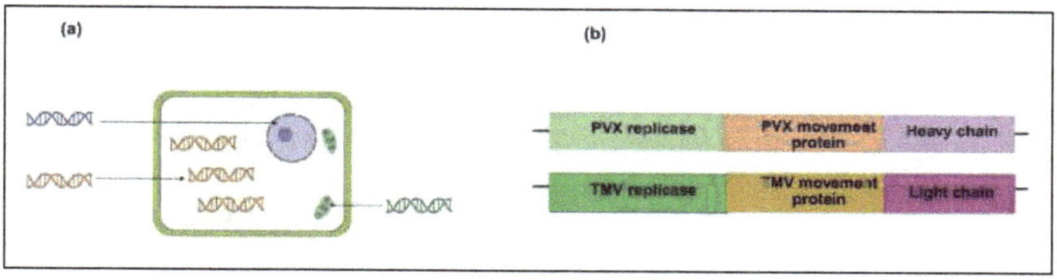

Fig. 6.36 Schematic diagram of (a) a plant cell showing localization of transient (orange) and stable (purple, green) expression systems (b) a portion of the two vectors used to produce full-length IgG antibodies in plants. Each vector carries the viral replicase and movement protein which drives the expression of antibody genes for the light or heavy chain. Regulatory elements are omitted from the diagram. The viruses used included *Tobacco mosaic virus* (TMV) and *Potato virus X* (PVX). The modified viral genomes were inserted into the T-DNA and delivered to plants via *Agrobacterium*. Because of the absence of the viral protein coats, both the viruses were unable to replicate

Poly (3-hydroxybutyric acid)

Plastics have a large number of important uses in modern everyday life. Plastics are polymers containing long-chain organic molecules with desired properties. However, synthetic plastics present serious environmental problems including disposal, persistence and pollution. Therefore, it would be beneficial to produce a biodegradable plastic-like polymer in plants. Poly (3-hydroxybutyric acid) or PHB, a type of plastic, which is naturally synthesized in many bacterial species, has been explored as an alternative to manufacture plastic products. A schematic diagram of PHB synthesis in the bacteria *Cupriavidus necator* (previously known as *Alcaligenes eutrophus*) is shown in Fig. 6.37a. With the aim of modifying plants to produce PHB, three genetic constructs, each containing one of the three PHB biosynthesis genes placed under the transcriptional control of the *Cauliflower mosaic virus* 35S promoter, including a chloroplast targeting peptide, were expressed in the model plant *Arabidopsis thaliana* (Fig. 6.37b). After transforming *A. thaliana* plants separately with each construct, two resulting transgenic plants, each with a different foreign gene integrated in its genomic DNA were crossed to generate a transgenic line with two of the PHB genes. Then the double-gene-transgenic line was selected and crossed with a transgenic plant line containing the third biosynthetic gene to combine all three PHB biosynthetic genes in the progeny (Fig. 6.37c, d). Using this approach PHB has been produced in leaf tissue at levels up to 4% of the leaf tissue's fresh weight. However, it should be noted that the plants were severely stunted. Nevertheless, this study provides an important first step of efforts to engineer the PHB biosynthetic pathway into larger plants for commercial purposes.

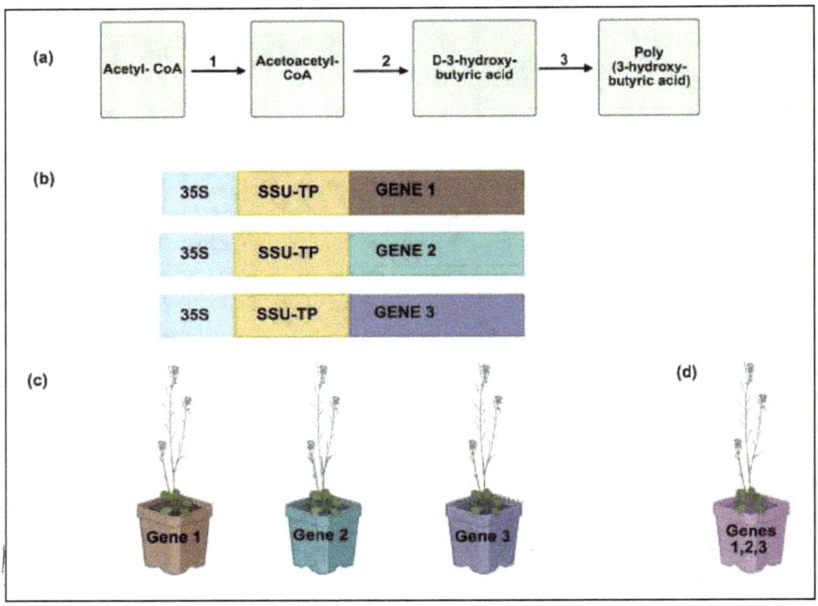

Fig. 6.37 Diagram illustrating **(a)** the pathway of poly (3-hydroxybutyric acid) synthesis in *Cupriavidus necator.* Enzymes are indicated by numbers: 1, 3-ketothiolase; 2, acetoacetyl-CoA reductase; and 3, poly (3-hydroxybutyric acid) synthase **(b)** three constructs, each showing a different gene for enzymes 1, 2, or 3; fused to the transit peptide of the small subunit (SSU) gene of ribulose-1,6-carboxylase (Rubisco). Each gene was driven by the *Cauliflower mosaic virus* 35S promoter **(c)** separate plants were transformed with each of the construct shown in panel b **(d)** independent transformant lines were crossed with one another until a transgenic plant line expressing all the three genes was produced

Edible vaccines

Using transgenic technology, it is possible to clone genes encoding proteins of microbial and viral infectious agents, and to express them in plants. The objective is that, by consuming plants containing an antigen as food, humans or animals can develop immunity to the pathogenic organism through the gut immune system. This method of invoking an immune response (Fig. 6.38) through oral delivery of plant-based subunit vaccine is an attractive alternative to injection, especially for developing countries. Advantages of edible vaccines include low production cost (as this approach eliminates expenses associated with production facilities, purification, sterilization, packaging, and transportation), safety (because they are needle free) and ease of administration (since vaccine administration does not require trained personnel). Additionally, as an edible vaccine is administered orally, the likelihood of developing mucosal immunity against pathogens that enter the body at mucosal sites is also increased.

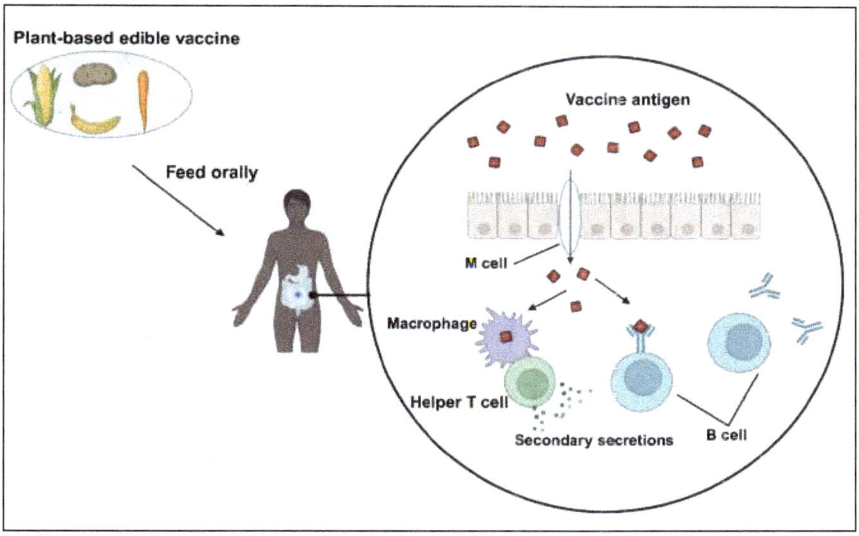

Fig. 6.38 Edible vaccines can elicit an immune response against an antigen from a disease-causing pathogen. The plant-derived antigenic protein, when ingested by the human host, binds to and is taken up by M cells in the intestinal lining and is then passed to other cells in the immune system, including macrophages and B cells. The macrophages present portions of the antigen to the helper T cells, which in turn secrete small molecules that activate B cells to produce antibodies that can neutralize the antigen

A number of edible vaccines are currently undergoing clinical trials (Table 6.9). Some of the diseases which these vaccines target include AIDS, cholera, tetanus, measles, malaria, gastroenteritis, diabetes, colorectal cancer, cervical cancer, and Alzheimer's disease.

Table 6.9 Plant-based oral vaccines and oral booster vaccines for infectious and non-infectious diseases

Antibody	Plant host	Target disease
Poliovirus VP1 antigen	Tobacco	Polio
VP40 protein	Tobacco	Ebola
Factor IX	Tobacco	Hemophilia A
α-glucosidase	Tobacco	Pompe disease
Cholera toxin-B-subunit	Rice	Cholera
GAD_{65}	Red beet	Type 1 diabetes
LTB-Syn chimeric protein	Tobacco	Parkinson's
Human β amyloid Aβ	Tomato	Alzheimer's
M-antigen	Lettuce	Measles
E7 protein	Algae	Human papillomavirus
Dengue viral protein	Tobacco	Dengue
Glycoprotein	Tomato	Rabies
Cry j1 and Cry j2	Rice	Japanese cedar pollen allergy
Envelope surface protein	Corn, potato, lupine, lettuce	Hepatitis B
Heat-labile toxin B-subunit	Tobacco, potato, maize	Enterohemorrhagic *E. coli*
Anthrax protective agent	Indian mustard	Anthrax
Capsid protein	Tobacco, potato	Norwalk virus

When oral vaccine development is pursued, one challenge is the degradation of the protein components in the gut before they elicit an immunogenic response. Hence, to be effective, higher levels of foreign protein expression in plant cells may be needed to elicit an immune response. Early work with plant-based subunit vaccines was reported using potatoes, due to their facile transformation. One of the ways to test if plant-derived antigens would function as an oral immunogen is by direct feeding of the raw materials to experimental animals. In one study, potato tubers were collected from plants transformed with fusion constructs in which the cholera toxin subunit B or A2 genes were fused to additional antigen genes. One fusion construct consisted of a 22-amino acid epitope from mouse rotavirus enterotoxin NSP4 fused to the C-terminal end of the cholera toxin subunit B protein, and the enterotoxigenic *E. coli* fimbrial colonization factor CFA/I was fused to the N-terminal end of the cholera toxin subunit A2 protein (Fig. 6.39). Mice fed with samples of transgenic tubers expressing both the fusion proteins developed an immune response against cholera toxin subunit B, mouse rotavirus enterotoxin NSP4, and *E. coli* fimbrial colonization factor CFA/I, and were protected against rotavirus-caused diarrhea. This study demonstrated the efficacy of edible tissue containing a recombinant immunogen to act as an oral immunogen when fed to mice.

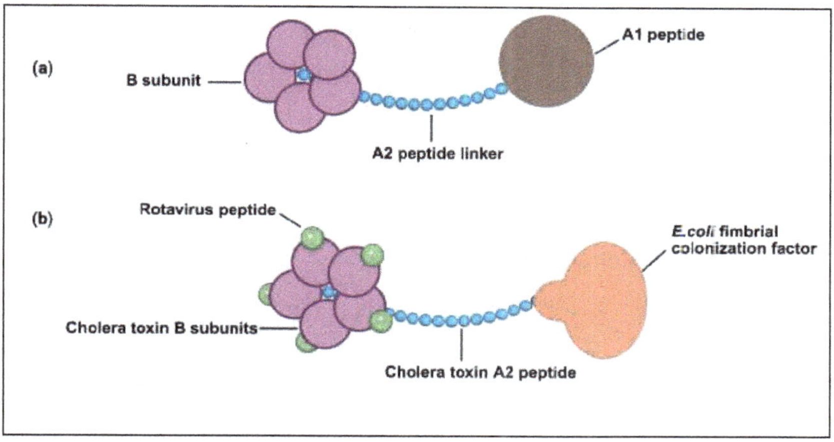

Fig. 6.39 A schematic showing an assembled (a) cholera toxin protein (b) an engineered cholera toxin B subunit (CTB) is synthesized without the A1 peptide. In the constructs, CTB is fused with a rotavirus peptide, and CTA2, which normally links A2 peptide to CTB, is fused to the enterotoxigenic *E. coli* fimbrial colonization factor. Both the engineered subunits are expressed in transgenic potatoes and reconstitute to form a cholera toxin-like protein structure

The ultimate goal is to develop a vaccine delivery plant system that could be directly used for immunization of children and adults. Plants that have been considered for this purpose include bananas, lettuce, carrots, legumes and cereal crops. Advantages of banana is that it is grown in many developing countries where vaccines are needed, can be consumed raw, and is eaten by both infants and adults. Although from a technical standpoint, banana trees require several years to mature. Legumes and cereals require cooking to be edible by humans, which would likely denature the proteins. Similarly, potato is cooked to make it palatable. On the other hand, carrot can be eaten raw, and is fast growing thus reducing the time needed between transforming the plants and the time when the tissues can be collected for testing and use. This might be a technical advantage.

Polio, short for poliomyelitis, is a highly infectious viral disease caused by the polio virus. Mass vaccination would be advantageous for populations in developing countries, where it is prevalent. However, the cost of vaccine production and distribution thwarts the achievement of this goal. A recombinant polio vaccine has been produced using the genes encoding the poliovirus capsid proteins VP1 and VP2, and this recombinant vaccine was expressed in carrot cells. The mice that were fed with transgenic carrot tissue developed an increased mucosal immune response compared to the mice that were fed with non-transgenic carrot tissue. Since the production system of carrot is available, one of the first biopharmaceuticals using this system (Elelyso™) for human use was introduced in the market for treatment of Gaucher's disease. Thus, oral delivery of vaccines using carrot cells can potentially be developed for commercial use.

6.8 Plant biotechnology in forestry

Natural forests provide economic and environmental benefit. As the global demand for renewable energy, paper, pulp, and construction materials from wood is growing rapidly, efforts have been made to expand plantation forestry. Genetic improvement of forest trees has been slow due to lengthy breeding cycles for trees, but biotechnological techniques have significantly improved tree breeding with this technology resulting in the transformation of a variety of tree species including Norway spruce, loblolly pine, radiata pine, Scots pine, Monterey pine, white spruce, black spruce, European larch, hybrid larch, and a wide range of eucalyptus and poplar species. Most of these trees were modified with traits such as enhanced growth rate, improved wood characteristics, pest and disease resistance, and various stress adaptations.

For the paper and pulp industry, a lower amount of lignin or lignin that is more easily extractable is a desired trait. Lignin represents about 15-35% of the wood biomass of various trees, and lignin reduces the efficiency of paper and pulp production. Delignification requires high energy, and generates large amounts of polluting chemicals. However, trees can be engineered to accumulate less lignin, to improve wood processing efficiency. This has been achieved by various means including the use of RNAi or antisense RNA to downregulate key enzymes in the lignin biosynthesis pathway (Fig. 6.40). The selected target enzymes were those that were located at an early stage of the lignin biosynthesis pathway. In an attempt to reduce the lignin content in quaking aspen (*Populus tremuloides*), one of the initial steps in the lignin biosynthesis pathway was altered. The enzyme 4-coumarate:coenzyme A ligase (4CL) catalyzes the conversion of 4-coumarate to 4-coumarate:coenzyme A, which acts as a precursor for downstream metabolites, including lignin and flavonoids. The 4CL enzyme exists in two isoforms-one is expressed in the epidermis of stems and leaves (involved in flavonoid biosynthesis) and the other is expressed in the developing xylem of the plant cell wall, which is where lignin normally deposits. The isoform specifically expressed in xylem tissues was cloned and its gene isolated. An antisense version of the gene was placed under the transcriptional control of the 35S promoter and introduced into aspen by *Agrobacterium*-mediated transformation. Compared with non-transformed poplar, the *Pt*4CL1 downregulated poplars were found to exhibit a 45% decrease in lignin content. Furthermore, the transgenic trees had a 15% increase in cellulose content and were about 25% larger (thicker stems, larger leaves and roots) than the non-transformed trees. In general, manipulating genes involved in lignin biosynthesis affects cell wall properties and, consequently compromises the tree's structural integrity or growth. However, in this case tree growth is not associated with lignin content.

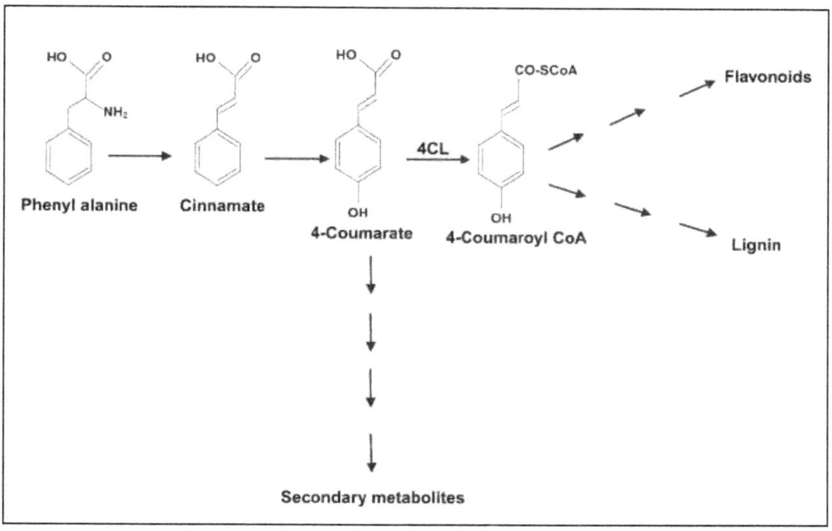

Fig. 6.40 Role of 4-coumarate: coenzyme A ligase (4CL) in the lignin biosynthesis pathway. By the substrate specificity of its isoforms, 4CL partitions metabolic flux between lignin biosynthesis and flavonoid biosynthesis. This directly influences how much lignin a plant produces versus other secondary metabolites

Lignin is composed of different monomers, mainly S-lignin (syringyl) and G-lignin (guaiacyl) (Fig. 6.41). The S/G ratio is an important determinant of wood quality and has been a target for genetic engineering. Wood products with a high S/G ratio are particularly desirable for biomass processing. The expression profile of the F5H gene, a member of the cytochrome P450 monooxygenase family, was studied in *Populus trichocarpa*, a fast-growing woody poplar plant. It was shown that the transcription factor, *pt*NF-YAA1 regulates the expression of the F5H gene in the lignin biosynthesis pathway, particularly the S-lignin synthesis pathway. This gene is highly expressed in the xylem cells, where lignin accumulates. Overexpression of *pt*NF-YAA1 led to an increase in the S-lignin content, while knocking out *pt*NF-YAA1 using CRISPR/Cas9 led to a decrease in the amount of S-lignin, as revealed in the biochemical and physiological analysis of transgenic poplars. When these two independent transgenic poplar lines were tested for the extractability of lignin during pretreatment, a step essential to make the cell wall carbohydrates more accessible to cellulose and hemicellulose degrading enzymes, the transgenic poplar which had a high S/G ratio, had the highest methane levels through anaerobic digestion. This suggests a more efficient conversion of biomass. This transgenic tree may reduce the need for harsh chemicals to remove lignin prior to enzymatic digestion of cellulose and hemicellulose. Moreover, poplar trees are commonly used as feed stocks for biofuel production, so it may be possible to apply this approach to more efficiently produce bioethanol.

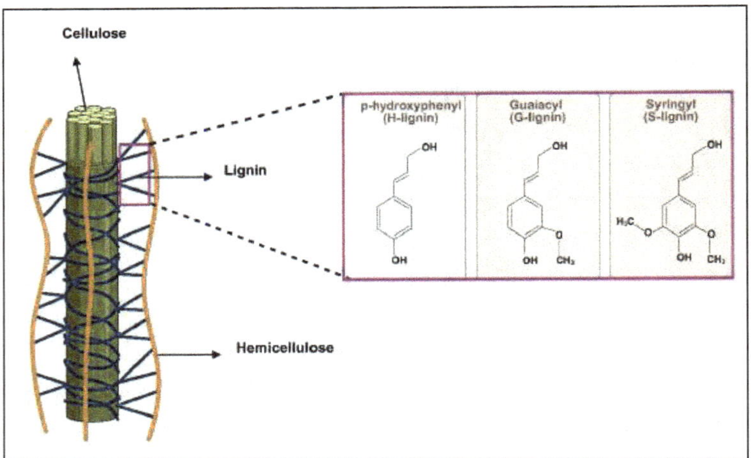

Fig. 6.41 Schematic representation of a plant cell wall. The structures of lignin monomeric units- p-hydroxyphenyl (H), guaiacyl (G), and syringyl (S) are also shown

6.9 Intellectual property and public acceptance

A patent (derived from the Latin word *patere,* meaning to lay open) is a legal document in which the government of a particular country (or group of countries such as the EU) gives the patent holder exclusive rights over the use, production, and sale of their invention. Moreover, depending on the claims of the patent that has been approved by a government agency (in the United States this is the Patent and Trademarks Office), the patent holder can develop other products that are directly derived from the original invention. Patent systems are not the same globally, and various countries have different rules. The duration of the ownership rights to an invention is typically 20 years from the date the patent application is filed, in all countries. While inventors themselves can draft the patent application, most inventors choose to work with patent attorneys. After a patent application has been submitted to the patent office, it is scrutinized by examiners during the examination process to ensure that it meets all of the established criteria for the patent to be granted. An inventor might seek patent protection internationally, for instance if they want to allow the production of the patented invention abroad. For any invention (product or process) to be eligible for patenting, it must meet several core requirements. First, it should be able to demonstrate novelty, which means, any invention that is filed for patent protection must not exist as another patent anywhere in the world. Second, the invention needs to be inventive. A patent cannot be granted for something that was merely, previously unknown. Rather, it must contain an inventive step that was not obvious to a "person skilled in the art." Third, the invention should be useful in some way, whether it is a technology, a gene sequence, a compound, a seed, or a plant. Products of nature are not patentable. Finally, the patent must have technical depth. In other words, the patent application should describe the invention in sufficient detail so that a skilled person can reproduce it. The applicant who was the first to invent is awarded the patent Therefore, while developing an invention it is important to keep dated notebooks detailing the subject matter such as a title, an abstract, background, summary of the invention (with figures or tables), industrial application, and, finally, a list of claims about the invention. Patents can be costly, which may provide an advantage to multinational corporations exploiting this approach. Many consumers worry about this as they feel that these large companies are the main beneficiary, possibly contributing to global financial disparities. This skepticism is often correlated with an increased perception of human or environmental health risk.

Biotechnology products are subject to specialized regulatory authorities. In the U.S.A, three agencies are involved- the U.S. Department of Agriculture (USDA), the Food and Drug Administration (FDA), and the Environmental Protection Agency (EPA) which focus on safety concerns, long-term effects from eating genetically modified (GM) food, labelling and product allergenicity. Yet, some argue that consuming GM foods might have unknown or harmful long-term health effects. Interestingly, there seems to be a regional difference in the relative perception of risks and benefits of GM foods. When Golden Rice was first introduced into the market it met with opposition; it took more than two decades before Golden Rice was approved for cultivation in countries like Philippines, Taiwan and Bangladesh. Public acceptance has been pivotal for policy support. One issue that stands out is that of labelling. GM labelling is supposed to give consumers the freedom of choice, but labelling GMO is not straightforward. It is simple to label a transgenic maize as such, but for example, maize products are used in varying amounts in thousands of processed foods, and it is difficult to label these in a meaningful and informative way. Similarly, does a product (like sugar, oil or starch) which is pure and free of DNA or protein, but is derived from GMO plant, still needs to be labelled as GMO? Despite their potential to generate food products, the public debate surrounding traditional genetic engineering is mostly based on sensational media coverage. Especially, in the early debates on GM foods, the media often used alarming phrases which fueled negative sentiments. Today, there is positivity surrounding gene editing techniques, such as CRISPR/Cas9, because such methods are perceived as more natural (they do not introduce any foreign genes into plants but rather modify existing plant genes). Nevertheless, there is still some uncertainty regarding the regulatory status of CRISPR-edited crops globally. Some jurisdictions such as Europe still restrict the use of CRISPR/Cas9 technology in plant breeding. Other countries including the United States, Canada, Australia and Japan have currently exempted most gene-edited crops from full environmental safety assessment and several gene edited products have been launched into the market without going through any formal regulatory review. It remains to be seen whether the European Union will adapt its regulatory frameworks to support the commercialization of edited plants.

6.10 Further readings

Akanmu, Akinlolu Olalekan, Michael Dare Asemoloye, Mario Andrea Marchisio, and Olubukola Oluranti Babalola. 2024. "Adoption of CRISPR-Cas for Crop Production: Present Status and Future Prospects." *PeerJ* 12: e17402. https://doi.org/10.7717/peerj.17402.

Adler, Reid G. 1984. "Biotechnology as an Intellectual Property." *Science* 224 (4647): 357–63. https://doi.org/10.1126/science.6584975.

Beyer, Peter, Salim Al-Babili, Xudong Ye, et al. 2002. "Golden Rice: Introducing the β-Carotene Biosynthesis Pathway into Rice Endosperm by Genetic Engineering to Defeat Vitamin A Deficiency." *The Journal of Nutrition* 132 (3): 506S-510S. https://doi.org/10.1093/jn/132.3.506S.

Bolaños-Martínez, Omayra C., Dania O. Govea-Alonso, Gladis Fragoso, Edda Sciutto, and Sergio Rosales-Mendoza. 2022. "Carrot Cells Expressing the VP1 and VP2 Poliovirus Proteins Effectively Elicited Mucosal Immunity." *Plant Cell, Tissue and Organ Culture (PCTOC)* 148 (3): 545–56. https://doi.org/10.1007/s11240-021-02205-0.

Butelli, Eugenio, Lucilla Titta, Marco Giorgio, et al. 2008. "Enrichment of Tomato Fruit with Health-Promoting Anthocyanins by Expression of Select Transcription Factors." *Nature Biotechnology* 26 (11): 1301–8. https://doi.org/10.1038/nbt.1506.

De Meester, Barbara, Ruben Vanholme, Thatiane Mota, and Wout Boerjan. 2022. "Lignin Engineering in Forest Trees: From Gene Discovery to Field Trials." *Plant Communications* 3 (6): 100465. https://doi.org/10.1016/j.xplc.2022.100465.

Devos, Yann, Pieter Maeseele, Dirk Reheul, Linda Van Speybroeck, and Danny De Waele. 2008. "Ethics in the Societal Debate on Genetically Modified Organisms: A (Re)Quest for Sense and Sensibility." *Journal of Agricultural and Environmental Ethics* 21 (1): 29–61. https://doi.org/10.1007/s10806-007-9057-6.

Falco, S.C., T. Guida, M. Locke, et al. 1995. "Transgenic Canola and Soybean Seeds with Increased Lysine." *Nature Biotechnology* 13 (6): 577–82. https://doi.org/10.1038/nbt0695-577.

Fedoroff, Nina V. 2015. "Food in a Future of 10 Billion." *Agriculture & Food Security* 4 (1): 11. https://doi.org/10.1186/s40066-015-0031-7.

Glick, Bernard R., and Cheryl L. Patten. 2022. *Molecular Biotechnology: Principles and Applications of Recombinant DNA*. Sixth edition. ASM Press.

Gupta, Mamta, Harish Kumar, and Sarvjeet Kaur. 2021. "Vegetative Insecticidal Protein (Vip): A Potential Contender from Bacillus Thuringiensis for Efficient Management of Various Detrimental Agricultural Pests." *Frontiers in Microbiology* 12 (May): 659736. https://doi.org/10.3389/fmicb.2021.659736.

Habben, Jeffrey E, and Brian A Larkins. 1995. "Genetic Modification of Seed Proteins." *Current Opinion in Biotechnology* 6 (2): 171–74. https://doi.org/10.1016/0958-1669(95)80027-1.

Hiatt, Andrew, Robert Caffferkey, and Katherine Bowdish. 1989. "Production of Antibodies in Transgenic Plants." *Nature* 342 (6245): 76–78. https://doi.org/10.1038/342076a0.

Hirschi, Kendal D. 2020. "Genetically Modified Plants: Nutritious, Sustainable, yet Underrated." *The Journal of Nutrition* 150 (10): 2628–34. https://doi.org/10.1093/jn/nxaa220.

Howard, Ted, and Jeremy Rifkin. 1977. *Who Should Play God? The Artificial Creation of Life and What It Means for the Future of the Human Race*. Delacorte Press.

Ikram, Aziz Ul, Muhammad Saad Shoaib Khan, Faisal Islam, et al. 2025. "All Roads Lead to Rome: Pathways to Engineering Disease Resistance in Plants." *Advanced Science* 12 (5): 2412223. https://doi.org/10.1002/advs.202412223.

Jin, Shuangxia, and Henry Daniell. 2015. "The Engineered Chloroplast Genome Just Got Smarter." *Trends in Plant Science* 20 (10): 622–40. https://doi.org/10.1016/j.tplants.2015.07.004.

Klińska-Bąchor, Sylwia, Kamil Demski, Yangmin Gong, and Antoni Banaś. 2024. "Metabolic Engineering of Omega-3 Long Chain Polyunsaturated Fatty Acids in Plants Using Different Δ6- and Δ5-Desaturases Co-Expressed with LPCAT from the Marine Diatom *Phaeodactylum Tricornutum*." *Scientific Reports* 14 (1): 9512. https://doi.org/10.1038/s41598-024-60141-3.

Kougan, Guy B., Turibio Tabopda, Victor Kuete, and Robert Verpoorte. 2013. "Simple Phenols, Phenolic Acids, and Related Esters from the Medicinal Plants of Africa." In *Medicinal Plant Research in Africa*. Elsevier. https://doi.org/10.1016/B978-0-12-405927-6.00006-0.

Kumari, Pritam, Poonam Jasrotia, Deepak Kumar, et al. 2022. "Biotechnological Approaches for Host Plant Resistance to Insect Pests." *Frontiers in Genetics* 13: 914029. https://doi.org/10.3389/fgene.2022.914029.

Lu, Haiwei, Guoliang Yuan, Steven H. Strauss, et al. 2020. "Reconfiguring Plant Metabolism for Biodegradable Plastic Production." *BioDesign Research* 2020: 9078303. https://doi.org/10.34133/2020/9078303.

Madhu, Alok Sharma, Amandeep Kaur, Deepika Antil, Sudhir P. Singh, and Santosh Kumar Upadhyay. 2023. "Plants as Bioreactors: An Overview." In *Plants as Bioreactors for Industrial Molecules*, 1st ed., edited by Santosh Kumar Upadhyay and Sudhir P. Singh. Wiley. https://doi.org/10.1002/9781119875116.ch1.

Mao, Ying-Bo, Wen-Juan Cai, Jia-Wei Wang, et al. 2007. "Silencing a Cotton Bollworm P450 Monooxygenase Gene by Plant-Mediated RNAi Impairs Larval Tolerance of Gossypol." *Nature Biotechnology* 25 (11): 1307–13. https://doi.org/10.1038/nbt1352.

Matin, Maima, Dalibor Hrg, Olena Litvinova, et al. 2024. "The Global Patent Landscape of Functional Food Innovation." *Nature Biotechnology* 42 (10): 1493–97. https://doi.org/10.1038/s41587-024-02410-0.

Muzaffar, Adnan, Sarfraz Kiani, Muhammad Azmat Ullah Khan, et al. 2015. "Chloroplast Localization of Cry1Ac and Cry2A Protein- an Alternative Way of Insect Control in Cotton." *Biological Research* 48 (1): 14. https://doi.org/10.1186/s40659-015-0005-z.

Menz, Jochen, Dominik Modrzejewski, Frank Hartung, Ralf Wilhelm, and Thorben Sprink. 2020. "Genome Edited Crops Touch the Market: A View on the Global Development and Regulatory Environment." *Frontiers in Plant Science* 11 (October): 586027. https://doi.org/10.3389/fpls.2020.586027.

Napier, Johnathan A. 2007. "The Production of Unusual Fatty Acids in Transgenic Plants." *Annual Review of Plant Biology* 58 (1): 295–319. https://doi.org/10.1146/annurev.arplant.58.032806.103811.

Narayanan, Zareen, and Bernard R. Glick. 2023. "Biotechnologically Engineered Plants." *Biology* 12 (4): 601. https://doi.org/10.3390/biology12040601.

Nehal, Nabya, and Priyanka Singh. 2024. "Biotechnology for Oil and Fat." In *Oils and Fats as Raw Materials for Industry*, 1st ed., edited by Divya Bajpai Tripathy, Anjali Gupta, Pooja Agarwal, Anuradha Mishra, and Arvind Kumar Jain. Wiley. https://doi.org/10.1002/9781119910558.ch2.

Oerke, E.-C. 2006. "Crop Losses to Pests." *The Journal of Agricultural Science* 144 (1): 31–43. https://doi.org/10.1017/S0021859605005708.

Pauls, K.P. 1995. "Plant Biotechnology for Crop Improvement." *Biotechnology Advances* 13 (4): 673–93. https://doi.org/10.1016/0734-9750(95)02010-1.

Sahai, Aastha, Anwar Shahzad, and Mohd. Shahid. 2013. "Plant Edible Vaccines: A Revolution in Vaccination." In *Recent Trends in Biotechnology and Therapeutic Applications of Medicinal Plants*, edited by Mohd. Shahid, Anwar Shahzad, Abida Malik, and Aastha Sahai. Springer Netherlands. https://doi.org/10.1007/978-94-007-6603-7_10.

Sánchez-León, Susana, Javier Gil-Humanes, Carmen V. Ozuna, et al. 2018. "Low-gluten, Nontransgenic Wheat Engineered with CRISPR/Cas9." *Plant Biotechnology Journal* 16 (4): 902–10. https://doi.org/10.1111/pbi.12837.

Sangeetha, Jeyabalan, Abdel Rahman Mohammad Said Al-Tawaha, Devarajan Thangadurai, et al. 2020. "Genetic Engineering for Developing Herbicide Resistance in Rice Crops." In *Rice Research for Quality Improvement: Genomics and Genetic Engineering*, edited by Aryadeep Roychoudhury. Springer Singapore. https://doi.org/10.1007/978-981-15-5337-0_10.

Shahid, Naila, and Henry Daniell. 2016. "Plant-based Oral Vaccines against Zoonotic and Non-zoonotic Diseases." *Plant Biotechnology Journal* 14 (11): 2079–99. https://doi.org/10.1111/pbi.12604.

Shepherd, Dionne N., Darren P. Martin, and Jennifer A. Thomson. 2009. "Transgenic Strategies for Developing Crops Resistant to Geminiviruses." *Plant Science* 176 (1): 1–11. https://doi.org/10.1016/j.plantsci.2008.08.011.

Shipman, Emma N., Jingwei Yu, Jiaqi Zhou, Karin Albornoz, and Diane M. Beckles. 2021. "Can Gene Editing Reduce Postharvest Waste and Loss of Fruit, Vegetables, and Ornamentals?" *Horticulture Research* 8 (1): 1. https://doi.org/10.1038/s41438-020-00428-4.

Smith, C. J. S., C. F. Watson, J. Ray, et al. 1988. "Antisense RNA Inhibition of Polygalacturonase Gene Expression in Transgenic Tomatoes." *Nature* 334 (6184): 724–26. https://doi.org/10.1038/334724a0.

Strobbe, Simon, Joshua Wesana, Dominique Van Der Straeten, and Hans De Steur. 2023. "Public Acceptance and Stakeholder Views of Gene Edited Foods: A Global Overview." *Trends in Biotechnology* 41 (6): 736–40. https://doi.org/10.1016/j.tibtech.2022.12.011.

Tang, Guangwen, Jian Qin, Gregory G Dolnikowski, Robert M Russell, and Michael A Grusak. 2009. "Golden Rice Is an Effective Source of Vitamin A." *The American Journal of Clinical Nutrition* 89 (6): 1776–83. https://doi.org/10.3945/ajcn.2008.27119.

Töpfer, Reinhard, Norbert Martini, and Jeff Schell. 1995. "Modification of Plant Lipid Synthesis." *Science* 268 (5211): 681–86. https://doi.org/10.1126/science.268.5211.681.

Vencill, William K., Robert L. Nichols, Theodore M. Webster, et al. 2012. "Herbicide Resistance: Toward an Understanding of Resistance Development and the Impact of Herbicide-Resistant Crops." *Weed Science* 60 (SP1): 2–30. https://doi.org/10.1614/WS-D-11-00206.1.

Visser, R. G. F., I. Somhorst, G. J. Kuipers, N. J. Ruys, W. J. Feenstra, and E. Jacobsen. 1991. "Inhibition of the Expression of the Gene for Granule-Bound Starch Synthase in Potato by Antisense Constructs." *Molecular and General Genetics MGG* 225 (2): 289–96. https://doi.org/10.1007/BF00269861.

Vo, Dang-Khoa, and Kieu The Loan Trinh. 2025. "Molecular Farming for Immunization: Current Advances and Future Prospects in Plant-Produced Vaccines." *Vaccines* 13 (2): 191. https://doi.org/10.3390/vaccines13020191.

Wang, Juli, Stacy D. Singer, and Guanqun Chen. 2024. "Biotechnological Advances in the Production of Unusual Fatty Acids in Transgenic Plants and Recombinant Microorganisms." *Biotechnology Advances* 76 (November): 108435. https://doi.org/10.1016/j.biotechadv.2024.108435.

Weselake, Randall J., David A. Fell, Xiaoyu Wang, Simon Scofield, Guanqun Chen, and John L. Harwood. 2024. "Increasing Oil Content in Brassica Oilseed Species." *Progress in Lipid Research* 96 (November): 101306. https://doi.org/10.1016/j.plipres.2024.101306.

Zarrelli, Armando. 2022. "Plants as Biofactories to Produce Food, Medicines, and Materials for a True Green Revolution." *International Journal of Molecular Sciences* 23 (10): 5827. https://doi.org/10.3390/ijms23105827.

Zhang, Yan, Jing Xu, Ruofan Li, Yanrui Ge, Yufei Li, and Ruili Li. 2023. "Plants' Response to Abiotic Stress: Mechanisms and Strategies." *International Journal of Molecular Sciences* 24 (13): 10915. https://doi.org/10.3390/ijms241310915.

Zhao, Yaling, Xin Yang, Guohui Zhou, and Tong Zhang. 2020. "Engineering Plant Virus Resistance: From RNA Silencing to Genome Editing Strategies." *Plant Biotechnology Journal* 18 (2): 328–36. https://doi.org/10.1111/pbi.13278.

Figures have been created using https://BioRender.com

www.ingramcontent.com/pod-product-compliance
Lightning Source LLC
LaVergne TN
LVHW052315200926
848481LV00002B/575
9798899662072